The
Pollen
Landscape

Joss Bartlett

Northern Bee Books

The Pollen Landscape

i Kathryn ac Elin Tesni

gyda chariad.

Published in the United Kingdom by

Northern Bee Books,

Scout Bottom Farm,

Mytholmroyd,

West Yorkshire HX7 5JS

Tel: 01422 882751

Fax: 01422 886157

www.northernbeebooks.co.uk

ISBN 978-1-912271-80-1

Design and artwork, DM Design and Print

Contents

List of figures.

1. Honey bees and pollen

My grandfather kept bees, in a couple of hives on the road verge behind his house. Like many beekeepers, he had his own ways: he believed, for example, that stings became less painful the more you had, and would put his hand into the hive in spring to get the first ones over with. Presumably the immunity faded over winter, because he had to repeat the process every year. And though most beekeepers concentrate on honey, and he won prizes for his in the village show, he lost interest in that and turned instead to making mead. I never tasted it, but I remember when he bottled it a scent so heavy it seemed to sink, and the colour when he opened it, like pouring yellow light into the glass.

Mead today is one of the less usual products of the hive, but in making it my grandfather was maintaining an ancient connection between people and bees. The Anglo-Saxons gathered in their meduærn, or mead hall, and in the early Welsh poem, the *Gododdin*, where the warriors feast for a year in preparation for battle, what they drink is 'medd'. The poem might be almost fifteen hundred years old, but the drink is certainly much older than that, as the similarity in the Welsh and English words may show. Sometimes languages have similar words because they inherit them from a parent tongue, and this is what has happened here. Words related to 'medd', like 'med' in Czech, and 'mádhu' in Sanskrit, occur in languages across Europe and south Asia, suggesting that the idea of a drink made from honey was already being expressed in the ancestral language from which all those tongues derive. Something like mead must have been enjoyed thousands of years ago.

The same is true of the honey from which mead is made. Related words can show a long history again: mêl in Welsh and meli in Greek, for example. And this time there is other evidence too - rock paintings in eastern Spain, dated to about 7000 BCE, show a human figure using a ladder to reach a nest, while bees fly around their head. In a world before cheap sugar, we can imagine risking stings for the sweetness in a comb of honey. Still, surely they would hesitate, and watch the bees for a while before they dared to rob them. And when they did so, these honey-gatherers must have noticed something much more obvious than the nectar from which honey is made: the balls of pollen that many of the bees would have carried on their legs.

Figure 1: pollen loads brought to the hive. The orange pellets are from dandelions, yellow from willow, white from wood anemones.

Perhaps they wondered about these coloured loads, or perhaps someone high up a ladder with angry bees around their head has other things on their mind. We don't know, because unlike honey, or the mead we can make from it, pollen has no everyday use to make it worth recording. There is no pollen in poetry, no paintings on walls, no reference to pollen loads at all until Aristotle, only four centuries BCE. And even after that it takes another two millenia and the invention of the microscope before anyone is sure of what they are. But in recent years, pollen has moved closer to centre stage. The relationship between insects and flowering plants, which has evolved over millions of years, may not be too far from a catastrophic collapse. The numbers and diversity of pollinating insects are declining, and there has never been more research into what they do. Pollen is part of that: the COLOSS honey bee research network, which has members in almost a hundred countries, carried out surveys in 2014 and 2015 on pollen brought to hives across Europe. In the UK, the National Honey Monitoring Scheme uses the latest methods of genetic sequencing to identify the pollen grains which end up in honey samples, and from those the plants from whose nectar that honey is made. Pollen today is a gateway to a dynamic and fascinating field of research. But it also offers something else for anyone who keeps or watches bees. Unlike the nectar or water that bees collect, but carry hidden away inside their bodies, with pollen we can see what they are doing. Sometimes it can be an indicator of what is happening in the hive: pollen collection in spring is a sign to the beekeeper that the queen has survived the winter, and the colony is raising young. Or if the colony has to rear a new queen, then a beekeeper will watch for pollen coming in to suggest that they have succeeded, and the queen is laying eggs. And more broadly it tells us about the world of the bees outside the hive: we can see from the pollen which plants they are visiting, and sometimes from that we can have an idea of where they have been. We can see how these things change with the seasons and from year to year, and with that, like anyone who finds a reason to stop and look, find ourselves connected more closely to the natural world.

Depending on the weather, the pollen year for the British honey bee can extend from the first pellets of hazel in January or February to the last loads of willow in October, a rhythm almost a full year long that can be followed with the pollen that they collect. The length of the season is a tribute to the adaptability of the honey bee. Bees which live in social groups, like the honey bee, probably evolved in a tropical zone and enjoyed a climate very different from the temperate regions where so many are kept now. 'It may be said that in our country,' grumbled R.O.B. Manley, a notable authority between the wars, ' the weather in summer-time is as a general rule about as bad as it can well be, considered from the point of view of the beekeeper'. Even so, the ability to store enough food to see them through winter allowed the honey bee to spread far to the north after the last ice age, and now their lives are regulated by seasons quite different from those that their ancestors would have known. The pollen year becomes a way to follow that, and even in my garden in north Wales, it can begin before January ends. Depending on winter conditions, the catkins of hazel are often advanced enough by then to release a puff of yellow pollen at the touch of a finger, and bees will work them on a sunny day. But usually the cold weather then closes in again, the catkins turn brown and have little to offer, and the bees retreat to their hives. So the real start of the year, the time when the bees show the first signs of the frantic, no time to lose attitude that they will keep for the next seven or eight months, comes in the second or third week in February. By then, there are clumps of snowdrops in the garden and along the base of the hedge in front of the house. If it is sunny enough to tempt the bees out to the snowdrops, then it is sunny enough for crocus flowers to open too, and soon each of those will have two or three bees circling the cup of petals, some gathering the orange pollen into a ball to be carried back to the hive.

But this early in the year, days suitable for foraging are likely to be the exception, and the start of the year is an uncertain time. In 2019, the year we will follow here, the bees were active throughout February and March. But a year before that they were busy on February 18th, and then a high pressure system settled over us, the wind blew from the north-east, and there was heavy snow. March remained cold and wet: I took one pollen sample on the 25th, but that was the only time in the month. Six or seven weeks of pollen were missed. For honey bees this far or further north, this kind of uncertainty can make spring the most perilous time. The honey they made last year to see them through the winter, and the pollen they stored too, may be running low, and poor weather can stop them collecting more. Meanwhile, inside the hive, the new season is well under way. The queen is laying eggs, there are new generations of bees to be made. And making bees, as we shall see, is what pollen is for. Just as much as the honey that we immediately associate with bees, the future of the colony is bound up with the pollen they can collect.

In this book, we are going to follow one colony of bees through their year of gathering pollen, to see what they find, and how they use it. We will look at the plants which make the pollen too, and the different aspects of the relationship that pollen creates between flower and bee. With the help of a microscope, we can examine something of the variety of pollen forms, and see how those can affect the bees as they collect them. The cycle of the hive is obviously bound to the seasons, because these help to determine what food they can find, but to the landscape as well, and the kinds of plants it provides. So we will look at where the bees of this north Wales hive find their pollen, and use that to consider how they cope, or struggle to cope, in the landscapes that they share with us, and that we alter and shape for our own needs. Altogether, as well as seeing the variety of pollen types gathered by a particular colony in a particular place, we are going to examine the whole relationship between honey bees and this part of their diet that they work so hard to find.

Aristotle's account of what we now call pollen loads is recorded in his *History of the Animals.* As well as honey, he wrote, 'they have another kind of food, which is called cerinthus (bee bread) which is of an inferior quality, and sweet like figs. They carry this upon their legs'. This he noted almost two and half thousand years ago. The knowledge wasn't lost: the Romans knew of 'bee bread' too, and every beekeeper since would have seen their bees bringing pollen loads to the hive, but investigating pollen isn't easy. Without the means to study it (or the inclination either, perhaps, when it offers nothing like honey or mead) for two thousand years after Aristotle little more was learned. Given this long silence, the first thing to consider is how we began to find out more.

2. Looking at pollen. February 2019.

In 1945, beside a hive near Cardiff, Mary Percival recorded the pollen collected by a single colony of bees. She watched the bees at 8 a.m. for 35 minutes, and repeated this every hour until half past five. Between April and August that year, she recorded almost half a million of the pollen loads that the bees brought in, and identified them on the basis of their colour and texture alone.

Dr Percival chose her method because she believed that direct observation of the bees gave the most accurate results. It certainly lets you see every pollen pellet that arrives, but it is hard to think of a more difficult way. She had to pick out each pollen-bearing bee through what is often dense traffic at the entrance to the hive. Sometimes, she noted herself, they arrived in such numbers that they were inside before she could count them. And she had to distinguish between shades of colour that can differ hardly at all. It would be like standing on Oxford Street on a Saturday afternoon, trying to record the colours of the shoppers' socks. And then giving the shoppers stings, and a suspicion of your plans. She did have one advantage, though: she was a university botanist with detailed knowledge of the flora around the hive. As part of her study, she spent one day a week examining the plants which grew within a half-mile radius of the hive, so that she had an idea of which kinds her bees might be able to find. So her choices for pollen identification would have been narrowed by knowing which plants were available at the time.

Not impossible, then, but even so the usual way of studying pollen is not to watch the bees directly but to use a pollen trap instead. Placed at the entrance to the hive so that the bees have to go through it on their way in, the trap is designed to part the pollen from the bee. Traps have drawbacks of their own, but as well as being easier than Percival's approach, they give us the pollen to examine at leisure. Then, as we shall see, we can use features other than the colour of the loads to find out what they are.

There is a market for pollen, and so there are commercial designs of pollen trap, but most people make their own. This means that they can differ in their details, but the principle behind them is usually the same. When the trap is set, the bees returning to the hive have to push their way in through a mesh, with holes about 5mm across. The holes let the bees in – with a struggle – but as they push their way through, the pollen balls are sometimes stripped from their legs. The lumps of pollen then fall through a smaller mesh (too small for the bees to get through to retrieve their load) and into a collecting tray further down. The bees find themselves in the hive without their pollen, perhaps like when we go into a room and can't remember what we wanted there.

Figure 2. Pollen trap

In this trap the collecting mesh is vertical. The trap is set by pushing the mesh into position, and a collecting tray is then left in a lower drawer. Other designs have horizontal mesh, with the trap at the bottom or the top of the hive, and some make the bees go through two layers of mesh instead of one, with the idea of making it more likely that the pollen will be caught. These kinds are often based on a design introduced by the Ontario Agricultural College in 1962, and look something like this one.

Drone escape

Figure 3: OAC type trap

One of the advantages of this kind is the drone escapes, fitted above the mesh. Drone bees are too big to fit through the pollen-stripping holes, and in their eagerness to leave the hive and look for a mate will either be crushed against the mesh, or maim themselves trying to get through. With the OAC design, they can find their way out through two larger exits and escape unharmed. Workers, of course, could also find their way back in through these and avoid the mesh, but the entrances are less obvious than the main opening, the bees are anxious to get in, and most go in through the trap.

No trap gets every load, which is a good thing: the bees need pollen, and we don't want to take it all. Mary Percival rejected a trap partly for this reason: worrying that her bees would starve. Some have tried to estimate how efficient the traps are, but the results vary. The British Columbia Department of Agriculture suggested that the OAC trap might take two-thirds of the pollen, and I have found similar results with this design. Others have reported traps intercepting almost half of what the bees brought in, but also cases where the catch was less than 10%. As for the effect on the colony, hives with pollen traps do seem to produce less honey. It may be that the reduced pollen harvest means that some of the foragers stop collecting nectar and turn to collecting pollen instead. This tells us that enough pollen is trapped to affect how the bees behave, but the colonies turn out to be no less likely to survive.

Another quirk of the trap is that some types of pollen can be missed altogether. Sometimes bees come back to the hive looking as though they can hardly lift their pollen loads through the air, and sometimes they carry so little that the pollen is not much more than a narrow sheath around their legs. These smaller loads are much more likely to go through the trap with the bee instead of ending up in the collecting tray, and species of plant from which only small loads are taken could be missed. Or if the bees collect only a few loads of a particular type, it may be that by chance all of those go through the trap uncaught. Either way, the traps probably have a bias towards the more common types of pollen, and some things may be completely missed. But hopefully we have most of the story, even if not every word.

A strong colony collects a good deal of pollen, and if the weather allowed my bees to fly, then the collecting tray was usually full when I came for it. It holds too many loads to examine each, so I then took a hundred of the pollen balls and divided them into groups of different colours. Repeating this three times produced an average membership for each of the colour groups. Under the microscope, I could then see whether a particular group was the pollen of one type of plant, or a mixture where the colours of the different kinds had been too similar for me to be sure which was which. If we look at a typical sample at the start of the year, I could separate the dark orange of snowdrop grains from the light orange of crocus, but then the light orange loads, examined under the microscope, turn out to be from not only crocus but coltsfoot (Tussilago farfara) too, a wild flower of waste ground and roadsides that blooms particularly early in the year. Another set of loads in the same sample were all similar shades of muddy white, but contained a mix of the pollen of four different plants. So a particular colour can represent a single species, but is often a mixture of different things, and then that mixture needs further work. With a 'pure' type like the snowdrops, the number of dark orange loads in my samples of a hundred is a direct estimate of the percentage of this species the bees have gathered. With the light orange loads, though, the two types have to be counted on the microscope slide to see how their number should be divided, in this case between coltsfoot and crocus. The process is repeated each day that the bees are collecting, and an average recorded for each plant species over the week. So the weather affects how many samples there are, but by the end of this year, 2019, there would be results for thirty five weeks.

All of this sampling – in the trap, in the tray and on the slide – means that, unlike Mary Percival, we won't have an exact count of the loads of, say, crocus, that the bees in this hive collect. Without a lot of repetition, we won't even have a particularly accurate percentage: it wouldn't make much sense, for example, to worry about whether the day's harvest is 25% snowdrop, or 26. Fortunately, it doesn't matter. We are interested in the general story of the year, and for this the ideal method already exists: categories to compare the contribution of each pollen type. The results from each week are used to place each type of pollen into one of four general groups, to give an idea of their relative importance at that time. Instead of trying to count exact numbers, we ask questions like: which kind of pollen is most important at a particular time ? For how many days or weeks is a particular type brought in ? Which types make only a minor contribution to the pollen economy of the hive and which do they rely on most ? Measuring pollen in categories like this was done first by the German scientist and beekeeper, Enoch Zander, in the 1930s. Zander is probably best known for having identified in 1909 the fungus which causes nosemosis in bees, but he was also among the first to examine the geographical and botanical origin of honey by looking at the pollen grains which the bees accidentally leave in it. Over a period of about twenty years, Zander was a key figure in creating this field of study, melissopalynology, which is still important today for checking whether the claims made on a label match the honey in a jar. A pollen type which made up more than 45% of a sample he termed *Leitpollen*, or leading pollen. Types between 16% and 45% were *Regleitpollen*, or accompanying pollen, and anything below 16% was *Einzelpollen* or odd pollen. A modified version has been in use since 1978, and under this scheme, pollen above 45% is called 'predominant' and pollen between 16% and 45% is 'secondary'. The difference is that the lowest class is now divided into two. 3-15% gives an 'important minor' pollen, and anything below 3% is 'minor'. These are the classes that we will use from now on to give a picture of what the bees collect. In 2019, it was the second week in February before they were active enough to make it worth setting the trap, and so that is where our pollen year begins:

February 2019, Week 2

Predominant	*Corylus avellana*	Hazel
Secondary	*Galanthus nivalis*	Snowdrop
Important Minor	*Viburnum tinus*	Lauristinus
	Alnus glutinosa	Alder
Minor	*Crocus*	
	Helleborus orientalis	Lenten rose

The results were similar in another year when February was mild enough for the bees to forage, and the comparison tells us that we probably have a good idea of the flowers that they can find at this time:

February 20ᵗʰ, 2017

Predominant		
Secondary	*Crocus*	
	Tussilago farfara	coltsfoot
Important Minor	*Galanthus nivalis*	snowdrop
	Alnus glutinosa	alder
	Corylus avellana	hazel
Minor	*Helleborus orientalis*	lenten rose
	Erica	winter heather

As we go through the year we will see that there is something unusual in these first results: a reliance on garden plants like snowdrop and crocus. The bees can even find one of the trees, hazel, on the edge of the garden where this hive is kept. Of all pollinating insects, honey bees form by far the largest colonies. A strong hive in summer might have 50,000 inhabitants or more, and building and maintaining a population of that size requires a lot of food, both nectar and pollen. This means that, most of the time, honey bees are industrial foragers, concentrating on places where there are plenty of flowers of the same type. The worker bees that collect the food are well-named, and they end their lives with a few weeks of intense effort, gathering food and bringing it back to the hive. They want fields, and what have been described as fields in the sky – the canopies of flowering trees. In a few weeks, for example, willow trees will bloom, and then the bees will scarcely bother with anything else and jostle their way through the hive entrance, laden with willow pollen in all shades of orange to yellow and green. But for now, the garden plants that they can largely ignore for most of the year are all that they have, and they bring back pollen from close to the hive, and from almost every plant which is in bloom.

The types of pollen collected at the start of the year show how colour alone can be an unreliable guide. As we have seen, some shades of orange are different, but others less distinct. Hazel pollen is yellow on the catkins, but more brown or white when the bees pack it on their legs, and loads of a similar colour come from other garden plants near to the hives – hellebores and winter-flowering heather *(Erica).* Pollen from the first of the trees that will be important over the next few months also appears – alder *(Alnus glutinosa)* and wych elm *(Ulmus glabra).* These can be plentiful sources. The small, red flowers of the elm look insignificant, but if they are picked and set in water in the house, a small heap of white pollen will collect around the vase. And they open in clusters at the tip of every twig, so that each tree offers rich forage for the bees. Alder can be important too, if there are patches near the colony of the wet ground that it favours. Bees visiting either tree bring back pollen loads of brown or a muddy white, hard to distinguish in the confusion at the entrance to the hive.

These various types, similar in colour, only show their differences under the microscope, and one is particularly beautiful there. This is the pollen of *Helleborus,* the Lenten rose whose many hybrids are common in gardens at this time of year. With this we can enter a world in which pollen study really begins, in the middle of the seventeenth century and the earliest days of the microscope as a scientific tool. This is the world of pollen morphology, the study of the many forms that pollen grains can take, and its importance to us is that it provides us with ways other than colour with which we can identify pollen types.

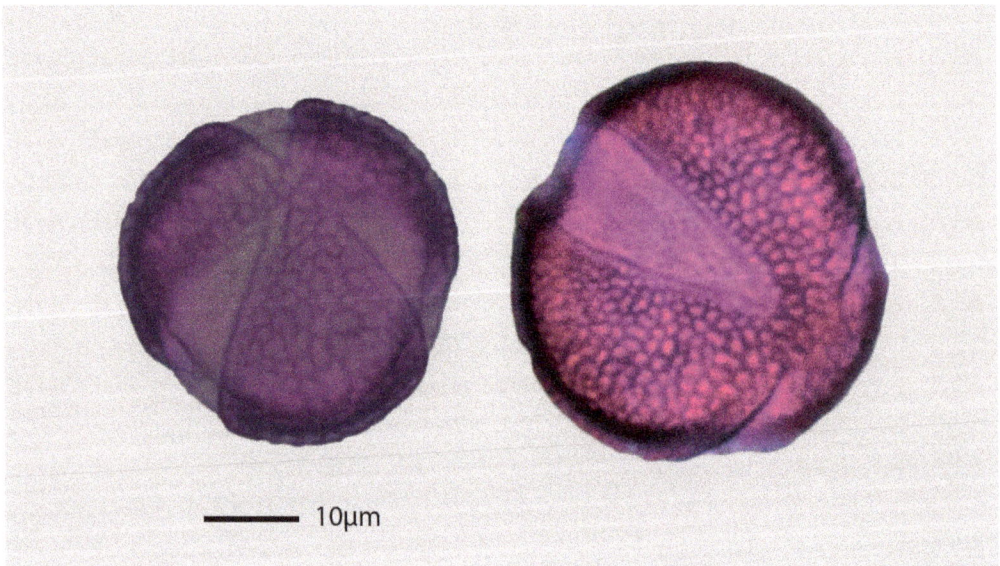

Figure 4: *Helleborus orientalis.* Family: Ranunculaceae (buttercups)

Each *Helleborus* grain is about 35 micrometres (µm) in diameter, which means that almost 30 of them could be packed on to a line a millimetre in length. It is spherical, as most types of pollen are, so we can think of it like a planet seen from space. In the picture on the left, we are looking down on one of the planet's poles, and the widest part of the image marks the equator. On the right, the grain is like a spinning top which has fallen on to its widest part, with the pole now pointing away to the lower right.

The size of the grains is one way in which different types of pollen can be distinguished from each other. The pollen of *Helleborus* is of fairly average size, but bees will collect the grains of evening primrose, which are almost four times bigger, and honey samples often contain the pollen of forget-me-not, only a quarter the size. Something so big or so small is distinctive enough, but there are hundreds of species whose pollen is in the same sort of size range, and then other differences are needed. One of these is the pattern of decoration on the surface of the grain. An electron microscope is needed to pick out every detail of this, but even with an ordinary light microscope, some features are usually clear. For *Helleborus*, the surface is reticulate – that is, as if a fishing net had been thrown across it. Compare that with the surface of the pollen of coltsfoot *(Tussilago farfara)*, one of the other types of pollen gathered early in the year:

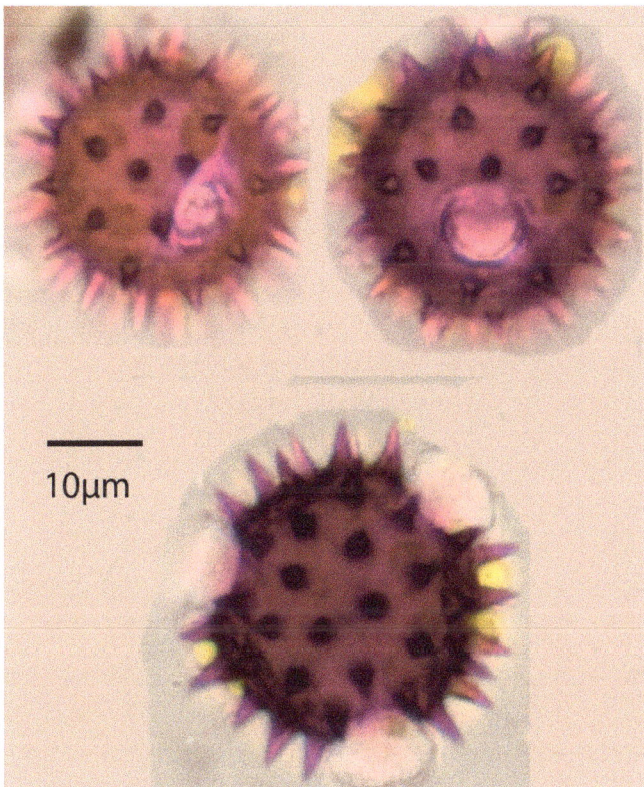

Figure 5: *Tussilago farfara.* Family: Asteraceae (daisies).

Here the surface is echinate – spiny – which means that it will catch on the feet and hairs of insects like bees, and be carried by them from plant to plant. Pollen transported on the wind is more likely to be smooth, reducing its resistance to the air. There are other types of surface decoration too, which we will see in the course of the year, and which can be clues about the kind of pollen we have found.

Once the shape, the size and the surface pattern of the grains have been noted, there is one more thing that the microscope can show. On the hellebore grain, the pattern of the net is interrupted by three paler areas, which widen as they run away from the poles down towards the equator of the grain. If this were a globe, they would be following the lines of longitude, and they are apertures – openings in the surface, the function of which we will see later on. For now, the important point is that they are another feature which can vary between different types of pollen, and so help us to identify which kind we have. The hellebore grain has the most common arrangement - three of these long, thin apertures – but there are many others. In some kinds, like red campion *(Silene dioica)* there are no long openings, but the surface net of the grains is perforated this time with the other aperture type: circular gaps called pores.

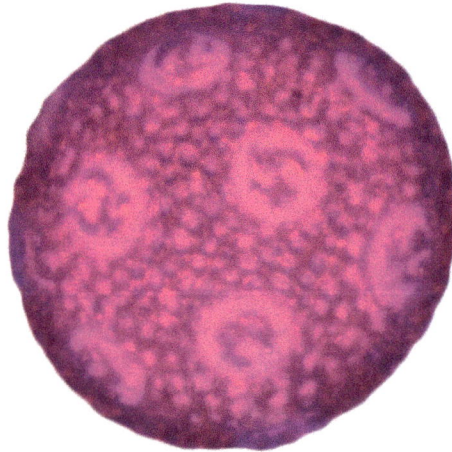

10μm

Figure 6: *Silene dioica*. Family: Caryophyllaceae (carnations).

Helleborus has its three long apertures, which are called furrows or colpi, as the only openings in the surface of the grain. Red campion has only pores. The two types of opening are also found in combination, as in these grains of black knapweed *(Centaurea nigra)* which the bees collect in August and July:

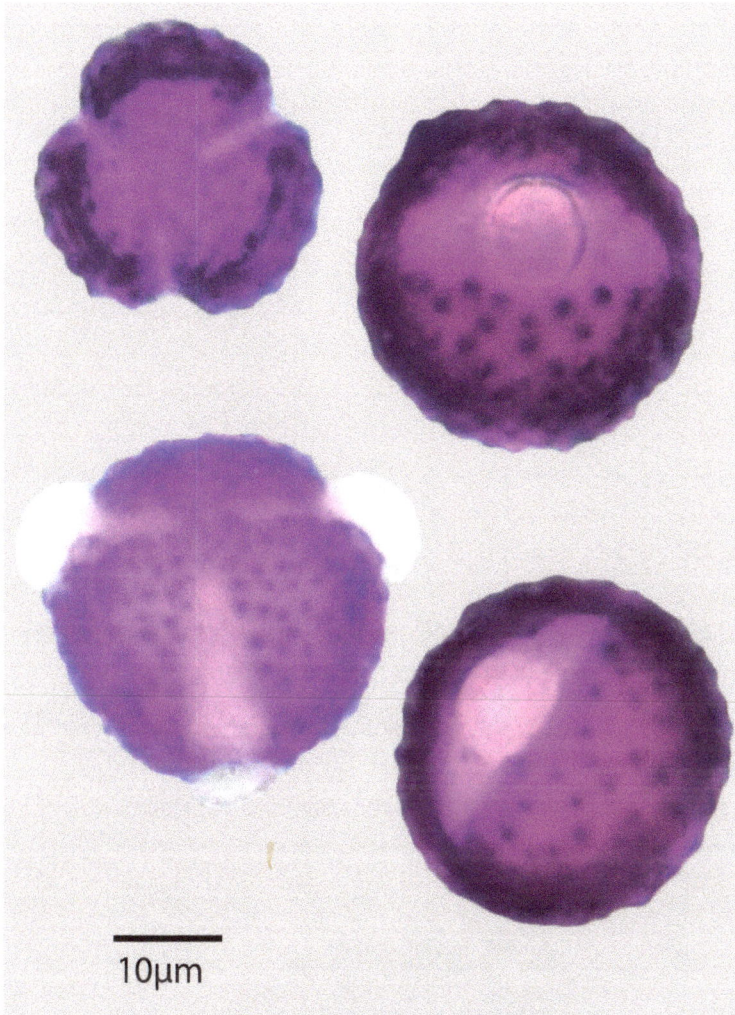

Figure 7: *Centaurea nigra*. Family Asteraceae (daisies)

Each circular pore is located in the middle of the long furrow. To go back to the image of the globe: if the furrows are lines of longitude, then their central point will lie on the equator, and in grains like knapweed, where pores and furrows are combined, the pores are placed around the equator of the grain.

Sometimes colour alone, or colour combined with season, can be enough to identify which types of pollen the bees are bringing home. The red of horse-chestnut or the late summer blue-black of rosebay willowherb can be picked out by standing beside the hive. But most types need confirmation under the microscope, using a combination of these characters of size, decoration and shape. So if we go back to the first sample of the year in February, we can look now in more detail at the three orange types. Coltsfoot has the spiny grains we have just seen. The bottom image shows how the ball of spines is interrupted by three apertures, and the top left grain how each of these is a furrow combined with a central pore. Snowdrop pollen grains are smooth, and under the microscope have a quite different shape, typically like the outline of a boat, as in the grain on the left:

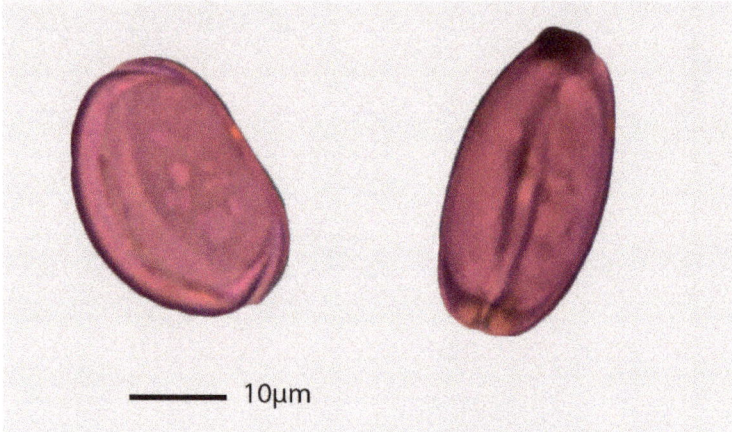

Figure 8: *Galanthus nivalis*. Family: Amaryllidaceae (amaryllis)

And where coltsfoot has three apertures, this time there is only a single furrow, running along the centre of the grain, as in the example on the right.

Crocus is the third orange type. It is spherical, like coltsfoot, but a giant of a grain, four times the average at about 100 μm across, and the spines on the surface are short, appearing in photographs like a sprinkling of black dots. The other grains in the picture are from snowdrop, and are a more typical size, showing how large crocus pollen is.

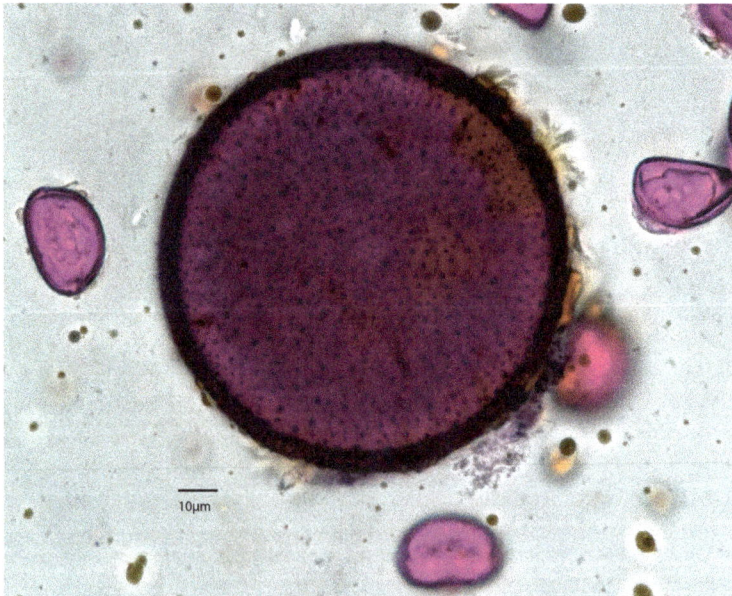

Figure 9: *Crocus*. Family: Iridaceae (iris)

We recognise pollen, then, as we do most things: with a combination of colour, shape, size, and any features that are unusual. The context is important too: bees can only collect from plants that grow in their district, and are flowering at the time. There are plenty of clues, but what makes it difficult is how small the grains are. Only the microscope reveals most of the differences between them and lets us go beyond trying to use colour to tell them apart. Its absence largely explains the long silence after Aristotle, and in the next chapter we will see how it allowed us to take the next steps in the study of pollen and bees.

The Pollen Landscape

3. Pollinator, pollination. March and April, 2019.

The exact origin of the light microscope is uncertain, but most would place it in the Netherlands, at the beginning of the 17th century. An English doctor, Nehemiah Grew, and Marcello Malpighi, an academic and physician in Italy, were then among the first to apply the new technology to the structure of plants, and in the process lay the foundations of what we now call palynology – the study of pollen and related spores. The first use of the microscope in relation to pollen, then, had nothing to do with bees, but was concerned with its role in the life of plants. The location of pollen in the male part of the flower was now described for the first time, although Grew used the term 'semet' for what we would today call the anther, the part of the flower where pollen develops, and from which it is released:

> These *Semets* (as I take leave to call them) have the appearance, especially in many *Flowers,* of so many little *seeds:* but are quite another kind of *Body.* For, upon enquiry, we find, that these *Semets,* though they seem to be solid, and for some time after their first formation, are entire; yet are they really hollow; and their side, or sides, which were at first entire, at length crack asunder: And that moreover the *Concave* of each *Semet* is not a meer vacuity, but fill'd up with a number of minute Particles, in form of a *Powder.* Which, though common to all *Semets,* yet in some, and particularly those of a *Tulip* or a *Lily,* being larger, is more distinctly observable.

For the first time here, in Grew's *The Anatomy of Plants* (1682), we find a powder of pollen located in the anthers, and what the new microscopes allowed us to see: that there were differences in the pollen of different types of plant. So, examining our own samples with a microscope, we can now separate the white-brown pollen loads of February, and they turn out to come from four different plants. We have seen one of them, *Helleborus,* already, but the most common is the pollen of alder. Like red campion, this has pores as its only openings, but there are far fewer, usually five arranged around the equator as in the bottom image, but sometimes only four. In the top row you can see the most striking feature of this species – the raised arches that link each pore, almost as if a mole had dug holes and then tunnelled between each, with a line of mounded earth to mark its trail.

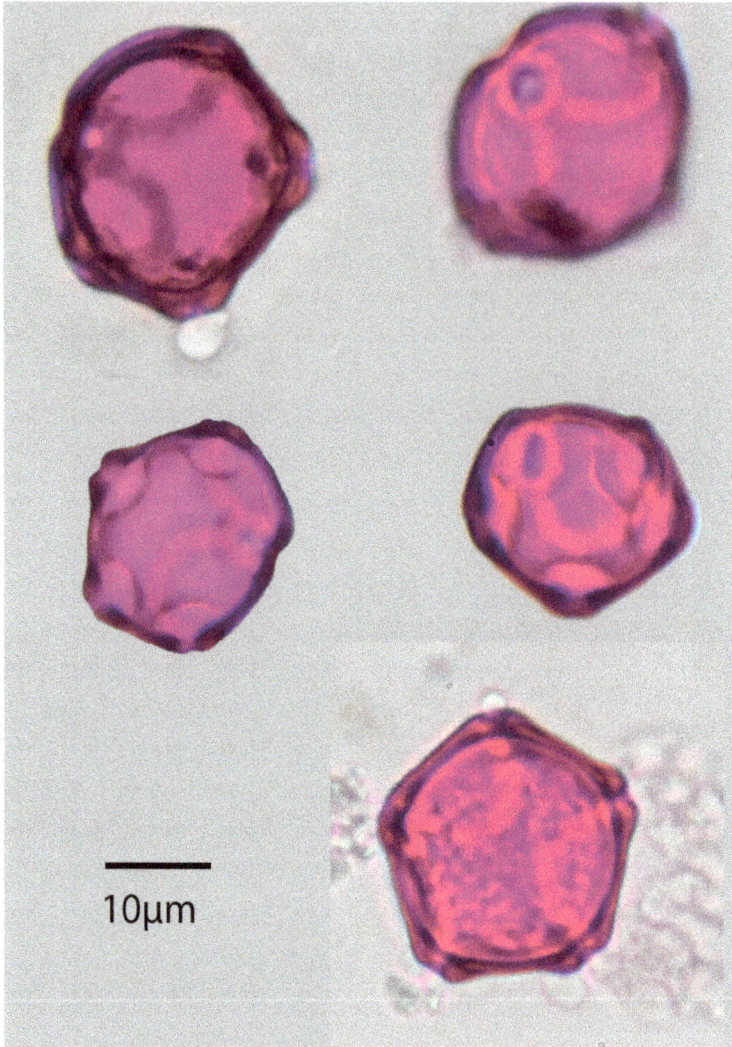

Figure 10: *Alnus glutinosa.* Family: Betulaceae (birch)

The third type is hazel. Hazel and alder belong to the same family of flowering plants, and both produce trailing catkins designed to release pollen on the wind. The chances of the wind blowing pollen to another plant of the same species are small compared with the precision of pollen delivered by insect, and so plants that rely on the wind tend to produce and release large numbers of grains.

More pollen is good for the bee colony, but from the point of view of the plant the insects are less than useful, because they are predators, collecting the pollen before it can be released. In February and early March, if the weather allows, hazel pollen is eagerly gathered by honey bees, and a tree on a sunny day can seem to have bees on every catkin. There are tiny red flowers on the tree as well, the female flowers waiting for pollen to arrive on the wind, but bees don't visit them, and the pollen that they collect is diverted from what the plant intended. Hazel pollen is similar to alder in size and colour, but quite distinct in shape. Alder is the unusual one: the rounded triangle shape of hazel is found in many hundreds of different species. In this case there is a pore at each apex of the triangle, and a ring around each pore.

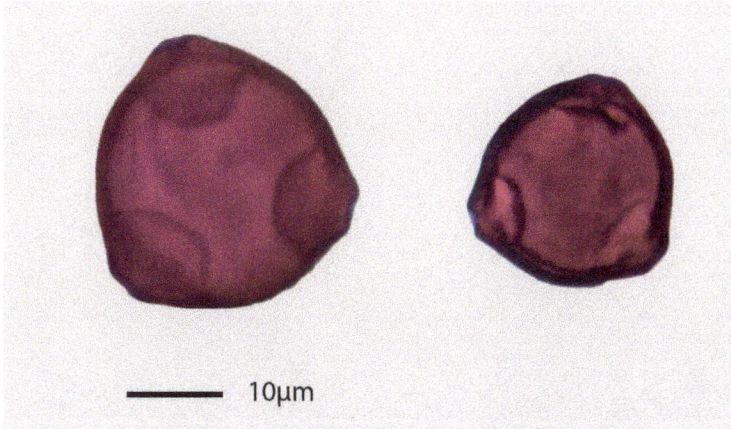

Figure 11: *Corylus avellana.* Family: Betulaceae (birch)

The large volume of pollen available from a tree is what the bees need to build up the colony for spring, but supplies can be scarce at this time of year, which is probably why garden flowers like snowdrops and hellebores are not ignored. Another garden plant that contributes to the earliest collections is heather. Late in the summer, of course, the heathers of high ground flower in such abundance that beehives are often moved on to moorland to make one of the few British monofloral honeys – honeys made almost entirely from the nectar of a single type of plant. Heather there is mainly ling, but the contribution now to early pollen comes from another member of the family: the winter-flowering varieties of *Erica* that offer garden colour when little else will bloom. Regardless of the type of heather, though, the pollen is the same, with its grains grouped in clumps of four, called tetrads.

Figure 12: *Erica sp.* Family: Ericaceae (heathers)

To extend Grew's description, we know now that pollen grains are formed from cells in the centre of the anthers, called pollen mother cells. Each of these divides to create four microspores, and each microspore will eventually develop into a grain of pollen. In most species, the four microspores separate before this occurs, so that we end up with the individual pollen grains that we usually see. But in the family of plants to which the heathers belong, the microspores remain together and develop as a group of four. This is what we have in the pollen of *Erica* in the first sample of the year, and will see again late in the summer when heather honey is made.

Grew left no description of heather pollen, but his microscope would probably already have been good enough to see the tetrad form. He, and Malpighi in his *Opera Omnia* of 1687, quickly discovered some of the features that we use still to describe pollen types. Differences in colour, of course, are clear without the microscope, but this may have been the first time that they were recorded. 'Some of these *Powders,*' noted Grew, 'are *yellow,* as in *Dogs-Mercury, Goats-Rue,* &c. and some of other Colours: But most of them I think are *white*'. Differences in size only the microscope could show:

Those in *Snap-dragon,* are of the smallest size I have seen; being no bigger through a good *Microscope,* than the least *Cheese-Mite* to the naked Eye. In *Plantain,* also through a *Glass,* like a *Scurvy-grass-seed.* In *Bears-foot,* like a *Mustard-seed.* In *Carnation,* like a *Turnip-seed.* In *Bindweed,* like a *Peper-Corn.*

All of these examples Grew recorded as being of a 'Globular Figure', but noted that the grains of other species had different shapes. Even some of the ornaments of the surface, like the spines of coltsfoot, were visible now – 'in *Mallow, Holyoak,* and all of that kind,' he recorded, 'they are beset round about with little *Thornes.*' Malpighi, meanwhile, described the single furrow that runs along the length of lily grains, and so made perhaps the first observation of a pollen aperture. Mapping out their new world, these pollen explorers added size, shape and apertures to colour as the characteristics by which pollen types could be known. They are the features we still use today.

So our pollen year has begun with seven types of pollen, and with these four characteristics, we can give them names. Some of these types would probably be brought back to almost every hive in Britain. Percival began her records a little later in the year, but at about the same time Dorothy Hodges was beginning a study of the colour of pollen loads, eventually published as a book of colour charts and beautiful pollen grain drawings in 1952. Her work involved a close study of the plants visited by bees around her home in Surrey, and she eventually recorded this as a calendar of the flowers they used throughout the year, based on her observations between 1940 and 1957. All the plants that we have seen so far were used there in February too. So few plants flower at this time of year that there is little room for variation, but some might change from place to place, or in the case of things like hellebores and winter heather, with gardening fashion. And some would differ with the weather from year to year. February of 2019 was unusually warm, and my bees brought in cherry pollen much earlier than they would normally do. Some of it I saw them bringing from a small tree close to the hives, of a type that flowers so early that it usually goes unvisited, but this year set fruit for the first time since I planted it, more than twenty years before. The popular winter-flowering viburnum, *Viburnum tinus,* also flowered well that year, the bees were able to visit it, and its pollen appeared in the first February sample, and the next one too:

February 2019, Week 3

Predominant		
Secondary	Alnus glutinosa	Alder
	Corylus avellana	Hazel
	Viburnum tinus	Lauristinus
Important Minor	Galanthus nivalis	Snowdrop
Minor	Crocus	
	Helleborus orientalis	Lenten rose
	Taxus baccata	Yew
	Ulmus	Elm
	Tussilago farfara	Coltsfoot

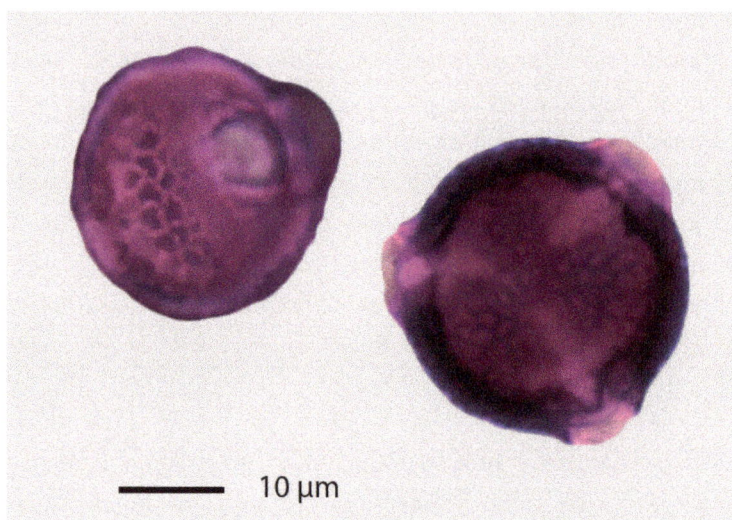

10 µm

Figure 13: *Viburnum tinus*. Family: Adoxaceae (moschatel)

Viburnum grains resemble those of the hellebore in colour and shape, and separating them shows how we rely on what the microscope reveals. There both have a surface pattern like a net, but we can see that the apertures of viburnum are a combination of furrow and pore, where the hellebore has only a furrow. There is an even smaller difference too: the furrows of the hellebore are stippled with spots, called granules, and those of viburnum are not. As microscopes improved, more and more types of pollen could be distinguished using details like these.

Often the weather now is too cold and wet for the bees to forage much. After that, they continued to collect from a range of sources, still using the few garden flowers available, and trees as they came into bloom. On March 9th, alder pollen made up more than half the loads brought in, but about 10% were elm, and a tiny amount of yew, two types not seen so far. 2018 was too cold for significant pollen foraging at all before the end of March, but in 2019 elm and yew were flowering well again by the second half of February, and their grains appear in the next samples.

There are two kinds of plants that make pollen. All of the examples so far have belonged to the group that produces its pollen in flowers – the angiosperms. The other group, gymnosperms, release pollen from cones, and yew is one of these. As you might expect from a plant so different, in evolutionary terms, from those seen so far, its pollen is quite distinct.

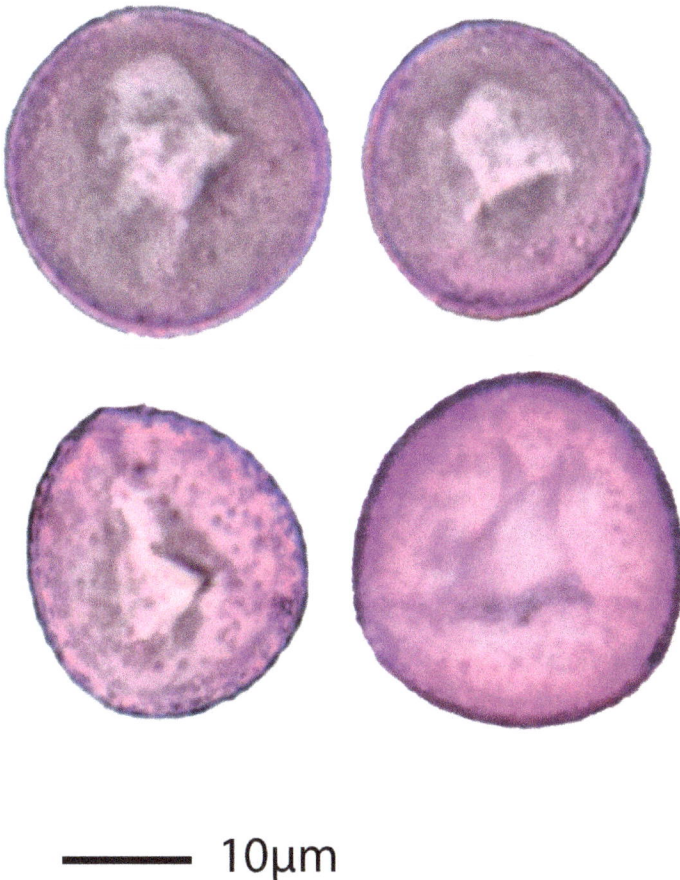

——— 10µm

Figure 14: *Taxus baccata.* Family: Taxaceae (yews)

It has a normal shape, and is small, but not unusually so. The peculiarity is that although it has no obvious pores or furrows, most grains look like those pictured, as though the surface has been torn. Only one other kind of gymnosperm pollen will be collected this year, and yew grains are like rare migratory birds: often absent altogether, but obvious when they do appear.

As for elms, there are several trees near to the apiary, and they flower profusely, if not for very long. The pollen is white and of average size; under the microscope it differs from the examples we have seen so far in the pattern on its surface.

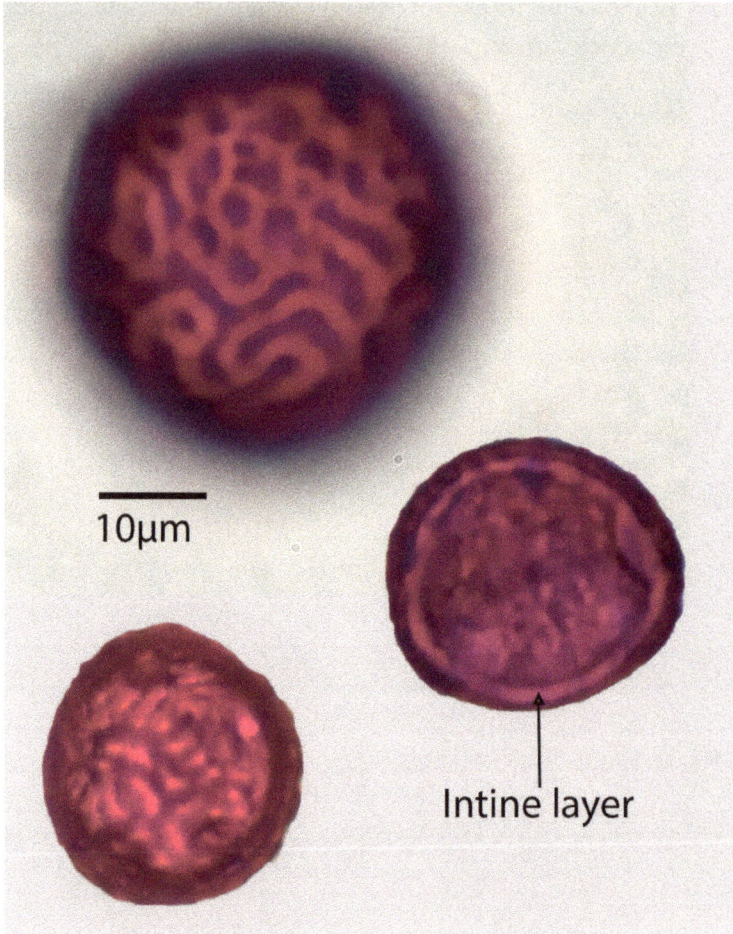

Figure 15: *Ulmus glabra.* Family: Ulmaceae (elms)

This type of arrangement, looking something like the folded surface of a brain, is called rugulate.

The surface pattern, the apertures, the colours and the shapes described by Malpighi and Grew are all formed by the various materials from which pollen grains are made. Some of these we will look at later on, when they will help us understand what pollen is for, but some are relevant now to the features that we use to identify pollen types. Like alder, the apertures of elm are in the form of pores, arranged around the equator, but in elm they are less obvious. An equatorial view of elm, however, does show how the outer wall of pollen grains consists of two distinct regions, and this is our first example of how the materials from which pollen is made create the features we can use to recognise it.

The German botanist Joseph Gottlieb Kölreuter described these layers in the 1760s, a century or so after Grew. Kölreuter, an apothecary's son from what is now southern Germany, was amongst the first to demonstrate the role of insects in carrying pollen between flowers, and so to move pollen studies on from being concerned only with plants. His experiments showed careful repetition and great patience. To prove that insect-pollinated plants could not be pollinated by the wind, he watched individual flowers from dawn until dusk, chasing away any insects that approached and showing that these protected flowers then set no seed. By then it was known that plants, like animals, exchanged male and female material to reproduce, and that pollen was an element in this. Even in plants like elm and alder, which bear both male and female components on the same individual, and which can therefore in principle fertilise themselves, Kölreuter realised that insect pollination still takes place. He showed too that in some plants the male and female parts mature at different times in order to prevent self-fertilisation from happening. With no discovery yet, though, of genetic material, Kölreuter could not know what fertilisation involved, and suggested that it was the result of mixing the oily contents of pollen with oily secretions from the stigma to which the pollen was delivered.

Much more detail was revealed with the invention of the electron microscope in the twentieth century, but the basic structure is as Kölreuter described it. The exine, the outermost wall of the pollen grain, is the layer that forms the spines or network or other ornaments of the surface, including the irregular folds of elm. The intine is an inner wall. It is particularly thick in elm grains , and can be seen as a distinct layer in a view focused on the equator.

The intine is built from carbohydrates, mainly cellulose and pectin, just like the walls of any plant cell. These walls are strong, and animals find them difficult to digest, which is why, for example, cows spend so much of their lives chewing what they eat. But they are fragile compared with the exine. This is made of sporopollenin, a material found only in pollen grains and the spores produced by some non-flowering plants. The name looks as though two words have collided, and they have. *Pollenin* was the term used in 1814 by a German chemist, Johann Friedrich John, to describe a chemically inert material which he obtained from tulip pollen. Then, when the French pharmacist and chemist Henri Braconnot examined the pollen of one of the grasslike sedges, he too found something inert and called it *sporonin*. The names were combined almost a century later. Even today, the exact structure of sporopollenin is not completely understood, but what is clear is that the inertness described by John and by Braconnot was the first discovery of a resistance to chemical attack and to decay that makes sporopollenin the toughest of natural polymers, once described as the diamond of the plant world. Pollen grains of elm like those the bees are collecting at the moment have been found buried deep in peat deposits, where they can be preserved for thousands of years and show us what plants were growing at the time when the pollen fell. Elm grains preserved like this have provoked discussion since the 1940s, when a sudden decline in the amount of elm pollen, dated to about 5000 years ago, was noted first. The same decline was then recorded from sites across north-west Europe, and has been attributed variously to climate change, woodland clearance for agriculture (itself partly shown by the appearance of pollen grains from cereal crops) and disease. In other locations an elm decline took place earlier, and the grains are evidence of events about 7000 years ago. Pollen can be archaeological evidence, hinting at lives far beyond written history. And, surviving because of its tough outer coat, it can take us even further back, as we shall see a little later on.

Writing in 1968 to show the readers of *Scientific American* the detail of pollen structure being revealed by the then-new scanning electron microscope, Patrick Echlin contrasted the extraordinary resilience of the pollen grain's coat with the brief life of the grain itself. If it seems strange to think of a pollen grain as having a life, then that may be because if we were asked about the function of pollen, we would be most likely to consider it as the equivalent of an animal's sperm cell: there to transfer the male part of the DNA, but without any independent life of its own. But it is more accurate to think of pollen grains in another way: as tiny, very simple, male plants. They travel, they germinate, and they grow, and the part that we usually think of: sperm cells, and the transfer of male DNA is only the very end of their lives.

The life of the grain, considered in this way as a microscopic male plant, begins as a microspore. One of our samples so far showed the grains of winter heather in their tetrads, clumps of four. As we saw, all types of pollen start this way, but usually the clumps separate into four distinct microspores. At this stage, each has a nucleus, containing the male DNA to be passed on to the next generation, and then this nucleus divides and the microspore becomes an immature pollen grain. It now has two cells, one inside the other. The inner cell is called the generative cell, and in many plants it remains unchanged until the pollen grain is transferred by wind or animal to the flower of another plant. Only then will the inner cell divide again to create two sperm cells, one of which will fertilise the egg. In other kinds of plant, the division into two sperm cells occurs before the pollen is transferred. In either case, the pollen grain lies dormant until transfer occurs. But once on the female part of a compatible plant, it absorbs water, and begins to grow. Now we can see the use of the apertures in the outer wall: they provide an entry point for water, and an exit point for the growing grain. Growth here is not the result of cell division, but rather the intine bulging out through one of the apertures and extending downward as a tube. The sperm cells are carried along as the tube grows and eventually delivered to the egg. The pollen grain, the tiny male plant, has produced sperm and performed its function as a male. Its brief life is over.

For our bees, it is still late winter. The pollen of only a few plants has completed its life cycle so early in our year. Mostly these are trees, with wind-dispersed pollen taking advantage of a clear flight path before it is closed by a canopy of leaves. In some years, our samples so far show that bees venture out too, and the pollen on these trees and some garden plants marks a new year for them as they do for us. In other years they huddle inside, perhaps for another month, suviving on what may be the last of their stores. But at some point in every spring the real pollen supply begins. Willows start to flower, and the bees forget almost everything else.

February 2019, Week 4

Predominant	*Salix*	Willow
Secondary		
Important Minor		
Minor	*Viburnum tinus* *Alnus glutinosa* *Tussilago farfara* *Prunus* group	Lauristinus Alder Coltsfoot Fruit trees

In March and April samples are often 80% or 90% willow pollen. There is a former landfill site near my apiary: the soil is waterlogged and there are dense patches of the trees there. The bees come to the hive with pollen in every shade of yellow, orange and green, and the colony's year has truly begun.

March 2019, Week 1

Predominant	*Salix*	Willow
Secondary		
Important Minor		
Minor	*Prunus* group *Alnus glutinosa*	Fruit trees Alder

The grains of willow are small, at about 18 microns across compared with the 25 – 30 of most types. They are round, and they have three apertures, each normally a furrow without an associated pore. Like the hellebores, the surface of each grain is decorated with a net.

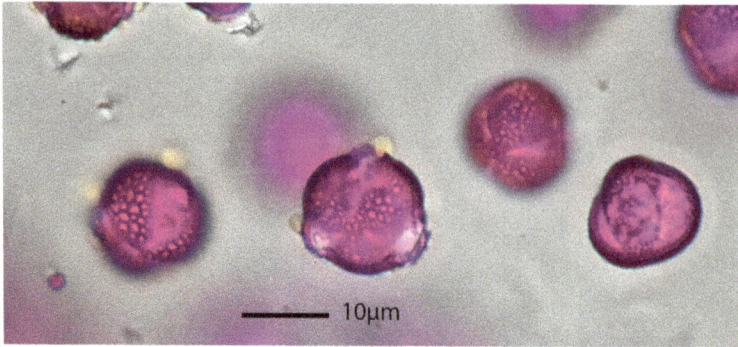

Figure 16: *Salix* sp. Family:Salicaceae (willows)

It has been said that beekeepers can do nothing more useful for their bees than plant a row of willows near the hives. The value of willow is not so much the duration of flowering – clover and bramble are used over a much longer season – as the timing. This is a harvest just when the bees need it most. Over the winter, the population of adult bees decreases, and is at its lowest point in early spring. But by then the queen has already begun to lay, and the workers which have survived the winter will have a growing number of larvae to feed. In 1937, C.L. Farrar, working for the U.S. Department of Agriculture, weighed a set of 12 hives, and subtracted the weight of the equipment to give a total weight of bees. Dividing these by the average weight of a single bee, he estimated that a colony could reach and maintain a population of 60,000 bees throughout the summer months. The large numbers are needed if the colony is to survive: more workers can gather enough stores for the next winter, and form a large enough cluster to stay warm through it. Large numbers also allow the colony to swarm, and have enough bees to share between the old population and a new one. Not all hives will contain this many bees (though some may have even more), but in every healthy colony, spring is a time for raising as many new bees as possible, and without pollen to feed the larvae, that can't be done. The willows flower at just the right time. And although they attract insects, they also seem to rely at least partly on the wind for their pollination. This means, as we have seen with hazel and alder, that they make a lot of grains. The time of year may explain this double strategy. A plant that flowers early in the season has the advantage that insects have few alternatives to visit, but the problem that the weather may be too bad for insects to be out at all. So using the wind may be a backup strategy for willows, in case the bees have to stay at home. As far as the bees are concerned, it means that willow pollen is not only timely, but plentiful too.

March 2019, Week 2

Predominant	Salix	Willow
Secondary		
Important Minor	*Buxus sempervirens*	Box
Minor	*Prunus* group	Fruit trees
	Taraxacum officinale	Dandelion
	Skimmia japonica	Skimmia
	Tussilago farfara	Coltsfoot
	Symphytum grandiflorum	Creeping Comfrey

The second March sample shows how the bees, needing large quantities of pollen, concentrate on the best sources. Apart from a temporary interest in some box shrubs which flowered well this year, the pollen collected was almost all *Salix*, with only small contributions from the other types brought in. Comfrey grows in large amounts all around the hives, and the bees collect a little of its pollen, often working alongside bumblebees once these emerge. But although the comfrey is so close, the greater effort needed to reach the willows, the nearest of which is at least 100 metres away, is clearly worth it. A study in the late 1970s, looking at the effectiveness of hives brought to an orchard to pollinate apples, found that willow was a significant competitor for the attention of the bees, which shows its attraction even next to other trees.

Although there is little of it, and there are years when none at all turns up in the trap, the pollen of comfrey is different from anything we have seen so far, and stands out amongst the mass of willow grains. Imagine a globe with the equator compressed, so that the two poles are pushed up and down, and you have the oval shape of a comfrey grain. Around the equator is arranged a neat row of 9-12 short furrows, each with a pore in the centre:

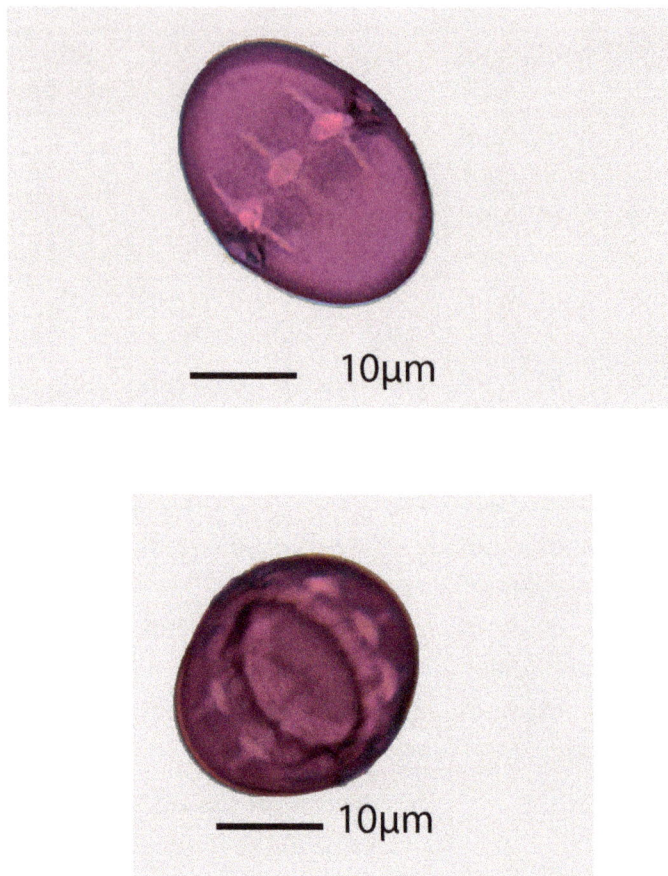

Figure 17: *Symphytum grandiflorum.* Family: Boraginaceae (forget-me-nots)

The lower image shows how the ring of apertures is arranged in a belt around the middle of the grain.

The materials from which pollen is made and which form their surface decoration are responsible for their colours too. Willow flowers are obvious at this time of year, creating splashes of yellow amongst the bare twigs of hedges or roadside trees. The catkins of hazel are also yellow, and in both cases the colour comes from the pollen grains. Although their vision is very different from ours, bees can see yellow too, but why have colour if the grains are only to be carried by the wind? One suggestion is that the pigments which create colour help to protect the DNA inside the grains. Pollen which is to be carried on the wind has to be held out in the open, where the wind can find it, and pollen in the open is exposed to UV rays that could damage its DNA. The yellow shades that are amongst the most common of pollen colours are formed by pigments called flavinoids, and they absorb the radiation that might otherwise reach the genes. At the same time, wavelengths at the red end of the spectrum are reflected, so that the pollen doesn't overheat. The first function of colour in pollen, then, was probably protective. But once insects appeared on the earth, the colour took on another use.

We think of insects being attracted by the bright petals of flowering plants, but pollen was the first colour signal that drew them in, and yellow the first colour they saw. Hoverflies will extend their feeding tubes towards yellow light even if they have never seen the colour before, a response that may go back to the first association between insects and flowering plants. Just as with petals, the colour of pollen is one of the things that attracts insects to flowers.

Other common shades of pollen grains, oranges and reds, are created by different groups of pigments such as the carotenoids. Some of these are yellow too, but of a more intense shade than the flavinoids. They probably also protect from radiation, but they are only found in pollen that is carried by insects, and their other function is as signals to draw the insect to the plant. These are the colours of the outermost layer of the grain. We have already looked at the tough exine which forms the pollen sculpture– the net and the spines and the other patterns – but plants which use insects to carry their pollen often cover this with still another layer: a sticky, fatty coat, first described in 1930 and named the *Pollenkitt.*

We saw orange pollen earlier in the year, from crocus flowers and snowdrops, but these are not native to Britain, and probably flower here most years before there are any insects about, other than the honey bees that we have also introduced. The earliest native that we have seen with these kinds of pigments is coltsfoot. Even then, experiments in Germany found that it flowered so early that it too had to rely on the honey bee as the main pollinator, rather than any native insect, but it also attracted small beetles with a combination of its yellow flowers and a sweet scent. Here the bees stop collecting coltsfoot pollen in March and it can be another month before they begin to bring back another bright orange pollen from a related plant, the dandelion. In 2019, though, with its unusually early spring, dandelion and coltsfoot pollen overlap.

Much less conspicuous, being a similar dull yellow-green to the flowers, is the pollen of box. There are some years when none is collected at all. It doesn't flower for long, and at this time of year poor weather can mean that the bees miss it altogether. But with some March sunshine, a shrub in a sheltered corner in front of the house is visited with enthusiasm, and then the pollen from the trap stands out for its unusual form.

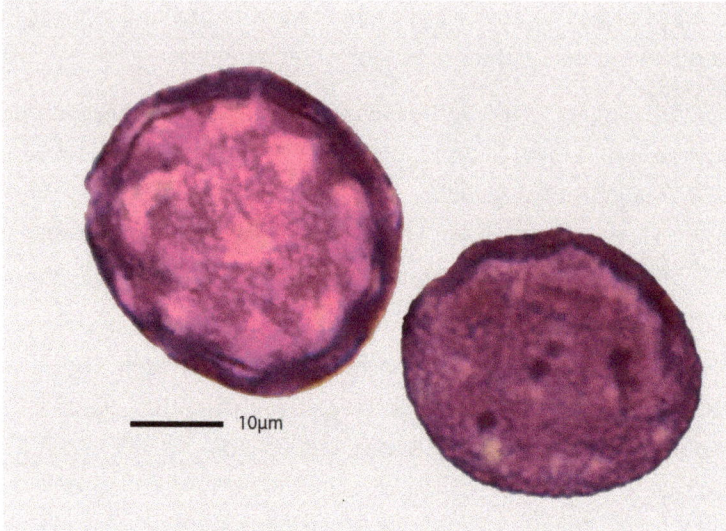

Figure 18: *Buxus sempervirens.* Family: Buxaceae.

Strictly speaking, it is of the same type as *Silene,* with a number of pores scattered across the surface. But there, as in other plants with this kind of pollen, the pores are neat, well-defined openings in the grain. In box, on the other hand, they are ragged, as if the surface has been torn.

March 2019, Week 3

Predominant	Salix	Willow
Secondary		
Important Minor	*Skimmia japonica*	Skimmia
Minor	*Prunus* group	Fruit trees

Another garden plant that a few bees visit, ignoring the general enthusiasm for willow, is *Skimmia.* These are widely-planted shrubs, and urban bees probably find them in most landscape planting schemes as well as in gardens. The pollen is red-brown, quite distinctive at this time of year, and the grains are something like those of comfrey, showing an arrangement of multiple furrows and pores around the equator:

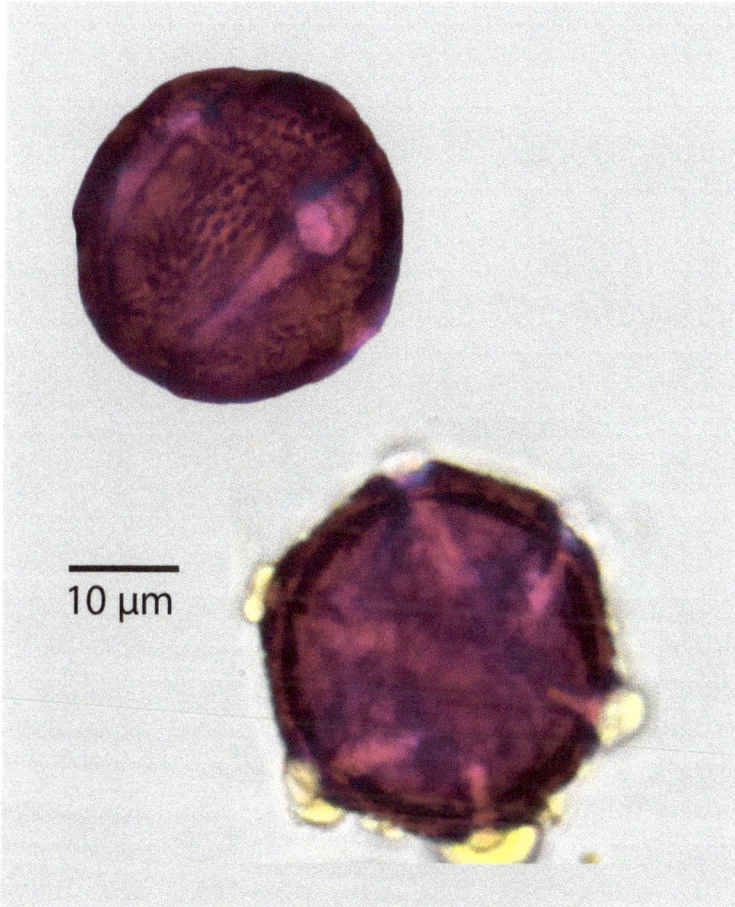

Figure 19: *Skimmia* sp. Family: Rutaceae (citrus)

This time, though, the polar view on the right shows that there are only six apertures instead of comfrey's nine or more.

March 2019, Week 4

Predominant	Salix	Willow
Secondary	*Prunus* group	Fruit trees
Important Minor	*Skimmia japonica*	Skimmia
Minor	*Taraxacum officinale*	Dandelion

A week further on, still a little earlier than normal, dandelions are in flower, and form the bright orange loads, much more orange than willow pollen, that appear over the next month and on into May. They show some resemblance to coltsfoot, to which they are related, in flowers and in pollen form. That had spiny grains, and dandelions also have spines, but now the apertures are so enlarged that from some angles there seems to be no more than a spiked scaffold protecting the inside of the grains.

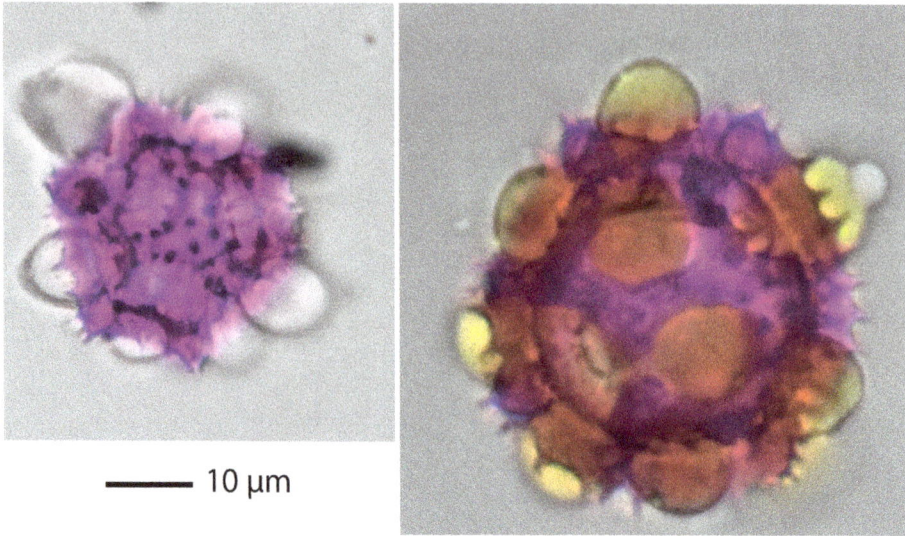

Figure 20: *Taraxacum officinale.* Family: Asteraceae (daisies)

One place where they flower now is on road verges, and council cuts mean that these are mown less often than they used to be, which extends the season. They quickly colonise any waste ground, and pasture fields that are not managed too intensively can be dotted with them. One reason for their rapid spread may be that they actually have no need of pollinators: dandelions can produce seeds without pollen transfer, although these will be genetically identical to their parents and miss out on any advantage gained from mixing genes with other plants. Even so, they attract many kinds of insects other than honey bees, and it is fortunate that something so common can be of such use. Here, at least: there are parts of the world where the dandelion is invasive and causes more problems than a few weeds in a lawn. We may even one day see fields where dandelions are the crop rather than a weed in the grass: there are plans to use them as an alternative source of rubber, and of inulin, a chemical that can be used as a food sweetener or in the manufacture of biodegradable plastic.

As we saw, the bright orange of dandelion pollen is partly created by pigments in the sticky coating that surrounds each grain. This *Pollenkitt* has many functions. It sticks to the insects which visit the plant, so that even creatures that have only visited for nectar are likely to carry pollen to their next stop. Once it gets there, the sticky coat then helps glue the grain to the stigma, where it can germinate and begin to grow. Sometimes the coat contains chemicals by which grain and stigma recognise each other and, if compatible, allow germination and tube growth to begin. This chemical handshake is one of the ways in which plants ensure that mating occurs mainly between male and female of the same species. Even if the 'wrong' pollen should germinate on the stigma, its pollen tube grows more slowly than the tube from pollen of the matching species, and is likely to lose the race to the egg. It may not always be wasted though: there is evidence that dandelion pollen which reaches the stigma of a different species may reduce the number of seeds that the other plant makes. Yet more chemicals in the pollen coat, it is proposed, cause this, and this partial crippling of its rivals may help explain why dandelions are so successful in the fight for space.

The coat glues pollen grains to one another as well, so that more are delivered. Wind dispersed pollens don't clump in this way: each travels separately, thousands of individual journeys compensating for the tiny chance each one has of success. But if your delivery is targeted, as with insect carriers, why not send several together, fertilise many eggs and create more than one seed ? In some plants, the arrival of more pollen grains can have just this effect, and this was the subject of more of Kölreuter's careful experiments. With *Hibiscus,* he reported, 50-60 pollen grains would result in the production of the 30 fertile seeds normally found in a seed capsule. There was plenty of pollen to achieve this: he had painstakingly counted the number of grains produced by a single flower and arrived at 4,863. But if he applied only 10 or 20 grains, then fewer seeds would result. Microscope technology at the time meant that Kölreuter couldn't know that more grains were needed because a single ovary in *Hibiscus* can contain about thirty eggs, but we can see now why arranging for many pollen grains to arrive at once can bring benefits to a plant.

Some chemicals in the *Pollenkitt,* then, convey messages to plants. Some create colour, and others signal to pollinators like bees through scent. Attraction to pollen odour has been known since the 1920s, and it may be that colour forms a long-range signal to guide insects, and scent provides information once the insect is nearby. This odour comes from the coat around the grains, and seems to result from oils similar to those that make flowers fragrant. Not only honey bees respond, of course: some solitary bees which are specific to certain flowers seem to use pollen odour to identify 'their' plants. Willows, still the main type that the bees are gathering now, show both kinds of insect message, but vary them between different types of plant. Willows are part of a plant minority, where male and female flowers grow on separate trees. The male plants have their bright yellow pollen to attract insects, where the female catkins are a less conspicuous green. Males also produce more scent. The effort that they put into attraction means that honey bees are more likely to visit male plants, and a bee that does go to a female willow will probably have been to a male first, and have pollen to transfer.

The male signal is what makes willow flowers obvious at this time of year. What we know of the visual system of bees suggests that willows stand out in the landscape to them as well, and they make full use of them, but then at other times they ignore what seem to us to also be prominent sources of food. Buttercups are a good example: later in the year the field in front of my apiary will be a sheet of yellow flowers, and the bees will ignore them completely. We will look at pollen choices later on, but a scent message may be how bees learn to avoid a pollen that can be harmful to them. If we compare the mixture of chemicals which create the odour of something like *Rosa rugosa,* a plant which uses a variety of insect visitors to transfer its pollen and provides abundant pollen for them, with the scent of the meadow buttercup, which feeds its insect visitors on nectar instead, they are strikingly different. To a bee, one smell may promise food, and the other danger.

One exception for the buttercup group of plants is the lesser celandine, at one time included in the same genus, *Ranunculus,* as the buttercups, but now called *Ficaria verna.* Its pollen is very similar to that of buttercups, and honey bees this year collected it in late March and early April. Perhaps, when the need of the colony is so great, and the sources of pollen so limited, the bees collect it because they must. Dorothy Hodges included celandine in her list of plants used by honey bees in early spring, and the dull orange pollen usually turns up in one or two samples around this time.

April 2019, Week 1

Predominant	*Prunus* group	Fruit trees
Secondary	*Salix*	Willow
Important Minor	*Skimmia*	
	Taraxacum officinale	Dandelion
Minor	*Ficaria verna*	Lesser Celandine
	Ulex europaeus	Gorse
	Hyacinthoides non-scripta	Bluebell

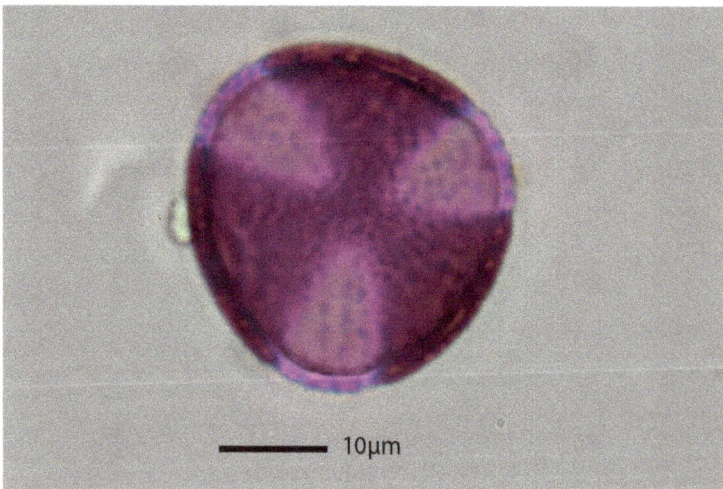

Figure 21: *Ficaria verna.* Family: Ranunculaceae (buttercups)

The grains are round, with three furrows and no pores. Each furrow is decorated with lines of granules, which at the equator look like stepping stones across the aperture's widest point.

Gorse can also be important at this time of year. It often flowers throughout the winter and uses colour and a coconut scent to attract mostly bumblebees and honey bees. Like dandelions, it has become an invasive nuisance outside its native European range, and its ability to flower at any time has probably helped it establish itself in novel parts of the world. It is never short of insect visitors, though it provides no nectar, and the bees only visit for its orange-brown pollen.

In its size, its shape of a three-cornered hat, and its three apertures, there is little to separate gorse from hundreds of other types of pollen. Under the microscope, the most striking feature is the way that the furrows resemble a long balloon, partly inflated, with a wide central region and two long, thin ends that almost meet at the poles. This region where the furrows approach one another is called the polar field, and in gorse pollen is distinctively small.

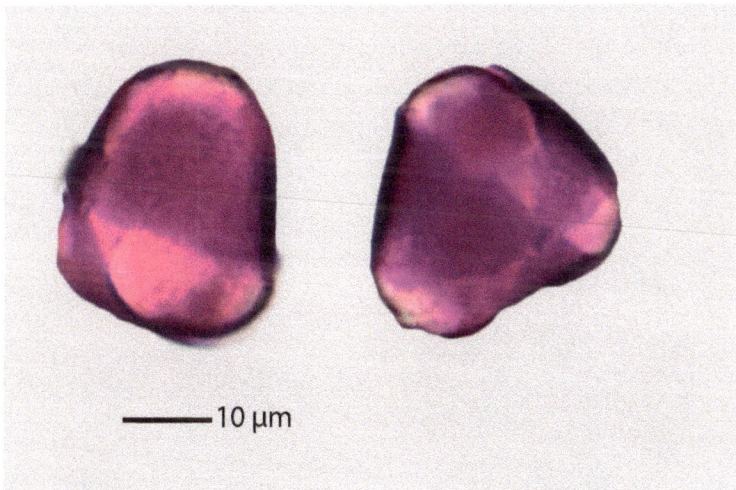

Figure 22: *Ulex europaeus.* Family: Fabaceae (peas and beans)

In some areas, gorse can be more plentiful than willow, and fill the same role of supplying pollen in large amounts early in the year. Any clump of gorse on a warm spring day is likely to hum with bees, but there are few such clumps close to my apiary, and gorse only contributes a little to the economy of its hives.

Another April flower which the bees use is the wood anemone, *Anemone nemorosa.*

April 2019, Week 2

Predominant	*Prunus* group	Fruit trees
Secondary		
Important Minor	*Salix*	Willow
	Taraxacum officinale	Dandelion
Minor	*Anemone nemorosa*	Wood anemone

The pollen loads which they collect from this are white and stand out amongst the pellets of willow grains. Like gorse, the anemone offers no nectar to its visitors, only pollen, and this seems to stand out bright yellow against the white petals. But this time the coloured pigments are in the anthers, and it is these rather than the pollen itself which advertises the reward. Anemones often grow together with the celandine, whose pollen the bees have been gathering for a while before the anemones flower, and the two are related in the evolutionary sense as well, both belonging in the Ranunculaceae, or buttercup family. One of the family traits they share is the appearance of their pollen: the surface pattern and the arrangement of apertures in the anemone resembles what we have seen already in celandine.

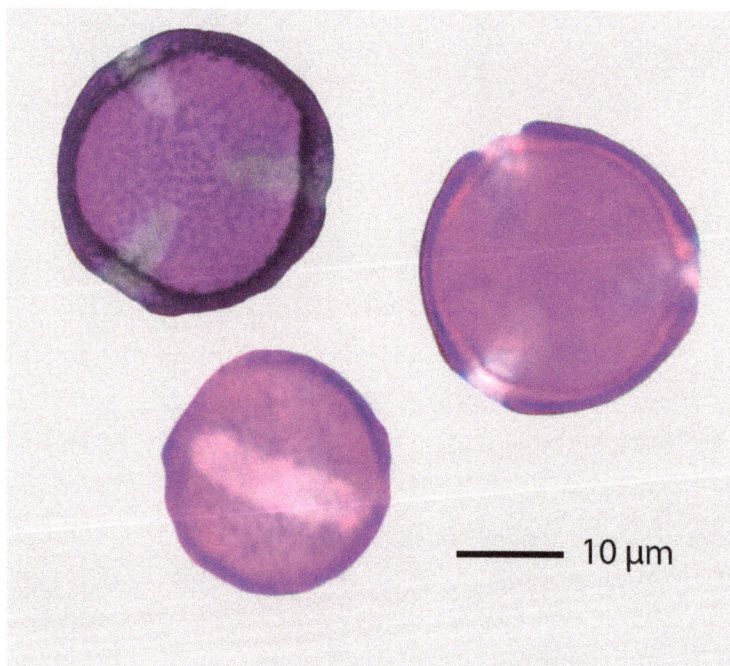

Figure 23: *Anemone nemorosa.* Family: Ranunculaceae (buttercups)

Similarities between the members of other plant families are easy to see as well. Nehemiah Grew had noted that spines were to be seen in '*Mallow, Holyoak,* and all of that kind', and as plants from all over the world began to be described and their family relationships explored, a possible role for pollen emerged. 'The figure of pollen I am inclined to think,' wrote Robert Brown in 1810, 'may be consulted with advantage in fixing our notions of the limits of genera.' By 1834 the German botanist, Hugo von Mohl, had for the first time produced a classification of plants by grouping, as he said, 'the species of each family according to the form of their Pollen.'

Pollen is usually a valuable guide in this way. We have already seen a resemblance between the grains of dandelions and coltsfoot, and when the various hawkbits and hawkweeds flower later in the year, we will find that their pollen can't be distinguished from that of their dandelion relatives at all. But sometimes, as Kölreuter noted, we can be misled, and quite unrelated plants can have similar pollen grains, or related kinds have pollen which hardly resemble each other at all.

So celandines, gorse and anemones have their stories to tell, but they are minor components in the pollen sample. The great change at this time of year, which in 2019 fell neatly on the line between March and April, is the shift from willow to fruit trees, and from trees which can at least partly use the wind to carry their pollen to those which rely entirely on insects to move it about. The fruit trees, mostly cherries, but with a mixture of apple and pear as well, depending on the district, dominate the pollen samples throughout the month.

April 2019, Week 3

Predominant		
Secondary	*Prunus* group *Salix* *Brassica*	Fruit trees Willow
Important Minor	*Hyacinthoides non-scripta*	Bluebell
Minor	*Taraxacum officinale*	Dandelion

The role of insects in pollinating plants like the fruit trees was not discovered until the 18th century, but as with pollen itself, Aristotle and his pupil Theophrastus were not far from finding it. They describe the process of caprification, where the fig crop is increased by hanging the branches of the male tree in female plants, or planting male trees close by the females which bear the fruit. This, they knew, meant that a small wasp would enter the fruit and so prevent it falling unripened. Herodotus, writing even before Aristotle, knew the process too, although he mistakenly ascribes it also to the date palm, which is actually pollinated by the wind. What neither understood was that the wasp had its effect on the fig by carrying pollen. Nor were they likely to: the fig and the wasp have one of the most specialised and unusual relationships of all the connections known between insect and plant, and as a place to begin to study insect pollination, could not have been much worse. The much more obvious transfer between honey bees and the flowers they visit went unnoticed. From our perspective, the pieces were there but unconnected, and so it continued for another two thousand years.

In 1750, Arthur Dobbs wrote to the Royal Society in London concerning observations he had made on his Irish estate. He was there, as he said, 'contemplating the Inhabitants of the Little World; particularly that most useful and industrious Society of Bees' and doing so partly because his 'View of doing Good, by making Discoveries of the Great World has been disappointed.' He was referring to his long dispute with the Hudson's Bay Company concerning the existence of the Northwest Passage. Believing that the Company was not trying hard enough to discover a northern route through to the Pacific, he had campaigned for their trade monopoly to be revoked, but in 1749 a Parliamentary inquiry decided against him. Dobbs retreated to his family estates, where his many interests included keeping bees, and there considered another of Aristotle's observations - that honey bees show flower constancy. That is, having visited, perhaps, a dandelion, they fly next to another dandelion, and another, ignoring flowers of other types along the way. This is not so much the case if they are gathering nectar, but Dobbs recorded that if they are collecting pollen, then their constancy is such that they will eventually return to the hive with a pollen load made up of grains of only one type. The loads that he examined supported this idea. Firstly, they were of uniform colour. Secondly, he observed, 'Bees, in the Height of the Season, return to their Hives with Loads of very different Magnitudes.' If bees collected pollen from a mixture of flowers, there would be no reason why they should not all return with loads of a similar size. But if a bee forages from a scarce plant, or one which produces little pollen, and then keeps to flowers of that type, it follows that it will come back with a smaller load. So as far as pollen foraging was concerned, Dobbs wrote, Aristotle had been correct. But by now Dobbs knew that plants were fertilised by pollen from one individual reaching the stigma of another, because this had been demonstrated fifty years before. So, he argued, the flower constancy of the honey bee might allow it to play a role in this process of pollination. If bees carried pollen at random, then 'like an unnatural Coition in the animal World, either no Generation would happen, or a monstrous one, or an Individual not capable of further Generation.' Providence, Dobbs concluded, had appointed the honey bee to go between flowers of the same type, carrying pollen from one to the next.

This is the task that our bees now begin. The willow catkins fade, and the bees turn to the insect-pollinated fruit trees that are their next main source of pollen. And here the principle that related plants have similar pollen types becomes a problem. So far, although we need a microscope, we have been able to distinguish between the various kinds in the sample. Even when, microscopically, anemone pollen resembled that of celandines, we could tell them apart with colour. But with the fruit tree family of plants, the Rosaceae, the members are so similar that it is particularly hard to say from which of them a particular pollen grain has come. It is as if, when we normally have no trouble distinguishing one sister from another, we came across a human family where we couldn't tell the cousins apart. The best we can usually do, without unusual expertise or an electron microscope, is to divide them into sub-groups based on size, or colour, or flowering time.

Most at the moment are species of *Prunus* – cherries or plums – and their pollen is various shades of green and brown. One of them, blackthorn, has been flowering in hedges for some time already, and is joined now by all the varieties of fruit or ornamental cherry grown in gardens or along our roads. Pear trees, which are species of *Pyrus* rather than *Prunus*, also flower at this time. Their pollen tends to be green rather than brown, but the colours can overlap and are not a reliable way to separate them. After thirty years of practice, a Somerset schoolteacher called Rex Sawyer published *Pollen Identification for Beekeepers,* and noted that in a mixed sample, all that could normally be done with this group was to lump them together as 'Prunus/Pyrus' type.

Pollen of this kind is slightly larger than the average, at about 30 – 40 micrometers. There are three furrows, each containing a pore at the equator, and the surface often shows wavy lines rather like the pattern of a fingerprint. This kind of surface pattern, different from any we have seen so far, is called 'striate'.

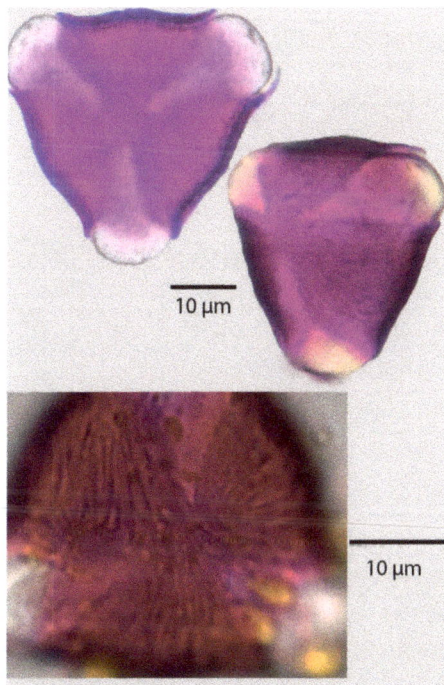

Figure 24: *Prunus* type. Family: Rosaceae.

One stimulus to bee research in the last century was concern about food supplies in time of war. In Britain, where blockades had focused attention on the need to boost home production, the Bee Department of the Rothamsted agricultural research station was established in 1944. Generous government funding sparked a productive period of bee research on the Hertfordshire estate, and over two summers there in 1945 and 1946, A.D. Synge recorded the types of pollen brought into the experimental hives. She set out her results, published in 1947, in an interesting way, summarising what was going on by dividing the year into distinct sections, each marked by some obvious change in the pollen collected. The first such boundary was the shift from willow to fruit trees, and Synge called this the change from early Spring to Spring. Later in the year, as we will see, her Spring period ended when the last significant tree pollen was finished, and the bees turned to herbaceous plants instead. Thinking about the pollen year in this way has two advantages. Firstly, it means that studies in different parts of the world, where the individual plants are different, can be compared by seeing whether there are still shifts between broad classes of plants like trees and herbaceous types. So in the 1990s, in Ireland, Mary Coffey and John Breen used Synge's divisions to summarise their pollen results, and in Wisconsin, in 1981, pollen collection was divided between tree sources from April to late May, shrubs from late May to mid June, and herbs until September. Secondly, the boundaries help us to iron out variations from year to year. In 2018 in my location in north Wales, *Prunus* pollen wasn't the main type until April 21st, three weeks later than in the following year. In 2020 the shift from *Salix* was a week or two later than the year before. The timing changes every year, but the event itself remains the same.

Dividing the pollen year into seasons: two examples.

Early Spring	Spring	Midsummer	Autumn	
Ash Poplar Elm	Fruit Trees Hawthorn	Clovers Sainfoin	Brassica Ivy	Synge (1947)

Early Spring	Spring	Midsummer	Autumn	
Willow Gorse Celandine Fruit Trees	Sycamore Beech Oak Hawthorn Ceanothus	Clovers Rosa Meadowsweet Elder Blackberry	Blackberry Heather Willowherb Ivy	Coffey & Breen (1997)

March	April	May	June	July	August	September

42

A super-sized version of this occurs in California, from which most of the world supply of almonds is harvested. Almonds, another kind of *Prunus,* flower there in February, and so much pollen is available and has to be transferred that an estimated 30 billion bees are moved there to carry out the job. Migratory beekeeping like this has never been anything like so significant in Britain, but where it does occur, tree fruits like cherries, plums, and particularly apples, are often involved. Apples flower a little later in the year, and their pollen, another with the same family characteristics, is probably mixed in with the samples then. In Kent or the West Country, it could be a significant type, and bees of economic value to the grower. A survey a few years ago found that about a quarter of commercial beekeepers moved hives to fields or orchards for pollination, and in the case of apples could add three or four thousand pounds to the value of the output per hectare. Other UK crops of interest to bee farmers were field beans, and oilseed rape, the pollen of which will appear soon.

The first significant *Prunus* sample from 2018 showed two other trees apart from the various fruit trees that the bees could use. Birch is widespread around the apiary, and, like alder and hazel, produces great quantities of pollen that are dispersed by the wind.

April 21st, 2018

Predominant	*Prunus* group	Cherry and plum
Secondary	*Salix*	Willows
Minor	*Taraxacum officinale*	Dandelion
	Symphytum grandiflorum	Comfrey
	Hyacinthoides non-scripta	Bluebell
	Betula pendula	Birch
	Fraxinus excelsior	Ash

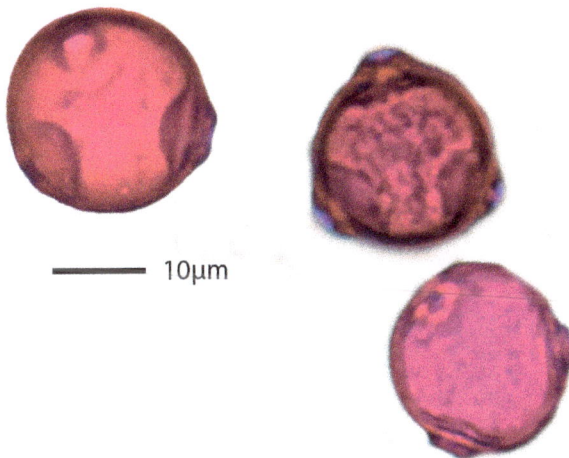

—— 10µm

Figure 25: *Betula pendula.* Family:Betulaceae (birches)

Birch pollen is a little larger than the grains of hazel, but otherwise they are similar, and both are in the same family. In the 1920s, William Burrell reported an analysis of pollen grains found in peat in the Pennines, and noted that under such conditions *Corylus* and *Betula* were indistinguishable. Burrell, better known for his work on mosses, was dipping his toes into palynology, a field of study then quite new in Britain. This science of pollen and other spores began in Sweden at the start of the 20th century, and was just becoming known in the German and English-reading world. In Sweden, Lennart van Post had found that a combination of waterlogged conditions, where the rate of decay is slow anyway, and the resistance of pollen grains to the process of decay, meant that the grains could remain in a recognisable state for many thousands of years. Now, by probing progressively deeper into the wet ground, and so further into the past, the pollen found could form a sequential record of the kinds of plants growing at the time. Not only this, but as van Post showed in the first example of a pollen diagram, published in 1916, the relative amounts of the different pollen types could also suggest how much of each type of plant there had been, and so reconstruct a picture of how the landscape would have looked. In this way, it was pollen that gave us our first tool for understanding how plants recolonised Britain after the last ice age, and how vegetation has changed over the millennia since. Birch, a plant which can survive harsher conditions than almost any deciduous tree, was amongst the first to grow and shed pollen here. In some places, grains from that time can still be extracted and recognised today.

Ash will deposit a thick layer of yellow pollen beneath a flowering stem if one is brought indoors and stood in water. With this plentiful supply and obvious flowers that open well before the leaves form, it might be sought out by bees if it flowered earlier in the year. They clearly prefer the fruit trees, though, and few loads of ash, or birch, are brought back to the hive; none at all in 2019.

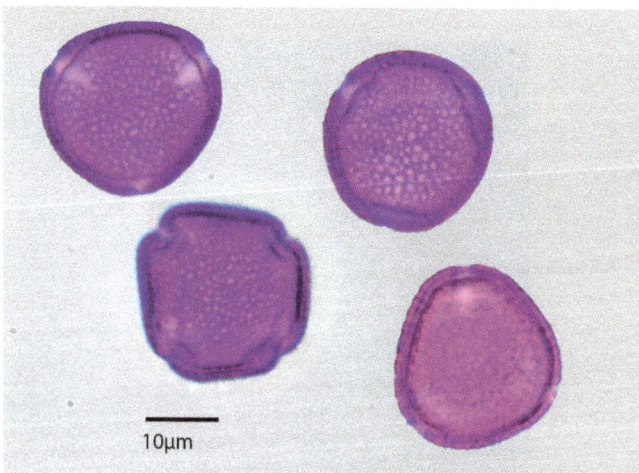

Figure 26: *Fraxinus excelsior* Family: Oleaceae (olive,lilac)

Throughout Synge's early Spring and Spring sections, the dominant pollen types are from trees. In general, as with ash, these flower before the leaves emerge: there is nothing then to block the wind taking the pollen from tree to tree. But the lack of leaves also means more light on the forest floor, and the anemones that we have seen already are one of the small plants that take advantage of that to flower early in the year. Another is the bluebell, earlier than ever in 2019, but usually towards the end of the month.

For my bees, there is only one nearby source: the woodland behind the hives. For a few weeks, just as the leaf canopy begins to form, the ground is the familiar sheet of blue, and you might expect it to hum with bees in the way that willow trees do, or cherries, but the pollen of bluebells is never their favourite kind. Even when the weather keeps them close to the hive, they seem to prefer the dandelions that are in flower at the same time. So although there are other woodlands within their range, it seems unlikely that they would go far for bluebells, and that the pollen they do collect is gathered close by.

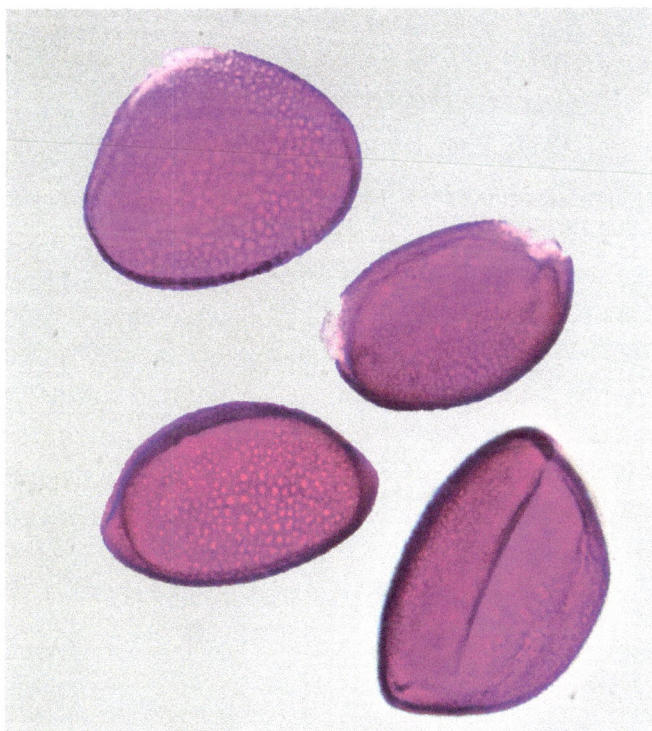

Figure 27: *Hyacinthoides non-scripta.* Family: Asparagaceae.

Bluebells were at one time classified with snowdrops in the lily family and have the typical pollen of that group, long grains where a single furrow forms the only opening. They often grow close to wild garlic *(Allium ursinum)*, and the two flower at much the same time in woodland where the drier slopes that bluebells favour meet flatter, damper ground where the garlic thrives. Although said to be a good nectar plant, and visited by bees for that, the pollen of wild garlic is rarely collected. When it does appear, in white-grey loads much like those of bluebells, it has the same lily-type shape, but is much smaller, about 30 µm long rather than the 50 of bluebell grains.

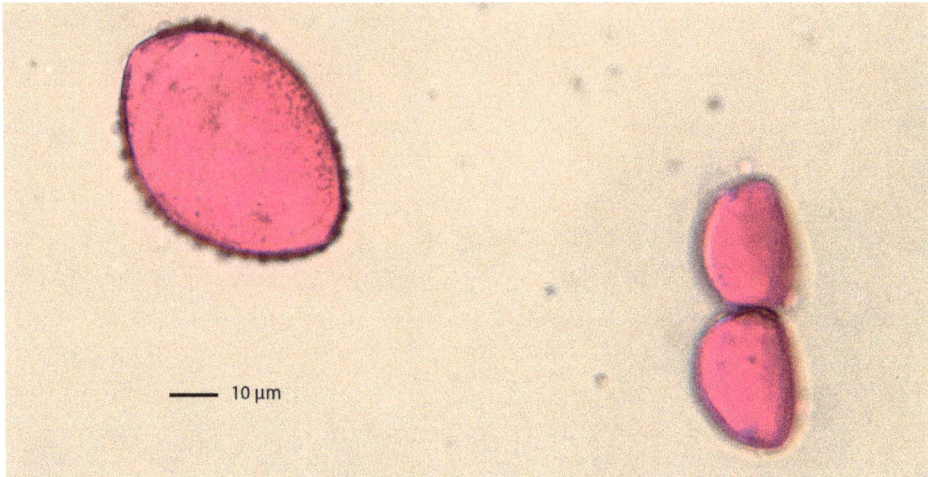

Figure 28: *Hyacinthoides non-scripta* (left) and *Allium ursinum* (right)

It was almost two hundred years after Dobbs wrote to the Royal Society about his bees before his contribution to our understanding of insect pollination was rediscovered. He himself had quickly tired of retirement and plunged back into the Great World. Some years earlier, with a consortium of like-minded landowners, he had bought 400,000 acres in North Carolina and encouraged its settlement by Scots-Irish emigrants. In 1753 he was appointed Governor there, and spent the rest of his life entangled in politics, war and the start of the American Revolution. He had one more botanical contribution to make, with the first written description of the Venus flytrap, but nothing further to say about bees. His letter was forgotten, and when the German botanists Hermann Müller and Paul Knuth published their great surveys of plant pollination in the last part of the 19th century, it was Dobbs' contemporary, Joseph Gottlieb Kölreuter, who was credited with the discovery of how insect visits transfer pollen from plant to plant.

April 2019, Week 4

Predominant	*Prunus* group	Fruit trees
Secondary	*Brassica*	
	Acer pseudoplatanus	Sycamore
Important Minor	*Salix*	Willow
	Taraxacum officinale	Dandelion
	Quercus	Oak
Minor	*Fagus sylvatica*	Beech
	Aesculus hippocastaneum	Horse chestnut

Alongside the cherries and plums in their orchards and gardens, forest trees other than ash and birch are releasing pollen - the hay fever season for those allergic to pollen from trees. A little after the cherries, some samples of beech pollen start to appear in the hive. Beech, oak and sycamore are more examples of wind-pollinated plants. They flower now, still before the canopy of leaves closes, and they release unimaginable amounts. A mature beech tree, for example, has both male and female flowers; each male flower when touched releases a small cloud of yellow dust and there are thousands of them on every tree. But although there are plenty of mature beeches close to the apiary, the bees stay far more interested in cherry trees and dandelions. When they do bring beech pollen back to the hive, it is in yellow loads, although the pollen of the copper-leaved form is much darker: orange or even brown.

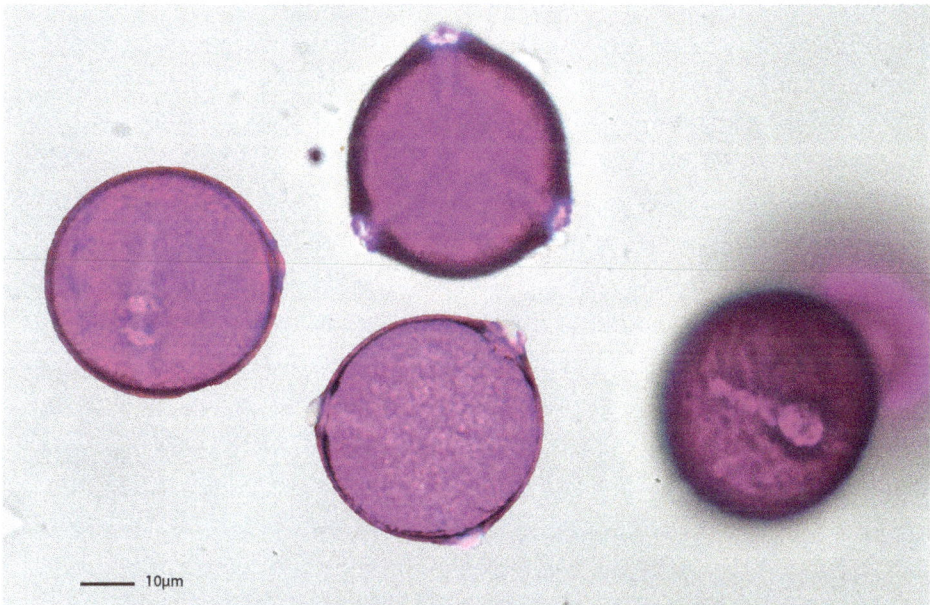

Figure 29: *Fagus sylvatica.* Family: Fagaceae (beeches)

Beech grains are larger than average, at almost 50 µm across, and the surface looks fuzzy, as though it would feel like touching a ball of wool. There are three pores, each set in the middle of a distinctively stubby furrow. Thinking back to gorse, with its long furrows and small polar field, this is the opposite. The furrows are short, so each ends a long way from the poles, and there is a large polar field into which the furrows don't reach.

Oak is even more common nearby, and is another wind-pollinated species that produces pollen in large amounts. For four weeks from mid-April to mid-May, it was the main type collected in a study in Wisconsin and provided most of the food the bees needed as they raised their young in spring. Here it appears over much the same period, but is never more than a minor component in what the bees collect. It is a dark or dull yellow colour, almost shading into orange, has three furrows without pores, and the rounded outline between the furrows produces something like the face of a cartoon cat.

—— 10μm

Figure 30: *Quercus*. Family:Fagaceae.

Nowadays we recognise Arthur Dobbs as part of the early understanding of insect pollination. There were others like him who for a while were forgotten too, such as Philip Miller, who wrote in 1721 that he had seen bees carrying pollen to tulips from which he had removed the stamens, and concluded that this was how these mutilated plants were then able to set seed. But Kölreuter it was who laid the firmest foundations for the subject. By the time of his death, after an academic career in Germany and Russia, there could be no doubt about the relationship between insects and flowering plants. He had begun to go further too, and to consider what we would now think of as the adaptations that connect the partners: what Darwin would come to see as the way that natural selection has shaped the structure and behaviour of insect and plant. It was on the basis of Kölreuter's work that his younger compatriot Christian Sprengel went on to present the structures of almost 500 species of flowering plants, many with adaptations for insect pollination, and the task of describing how different flowers are pollinated then continued for another hundred years.

From the first observations of Miller and Dobbs to the careful cataloguing of the German botanists, the 18th and 19th centuries saw the pollen connection between insects and plants made clear. For many plants, insects were their main or only pollen carriers, and complex relationships had sometimes evolved to connect the two. The benefit to the plant was clear as soon as it was known that pollen transfer allowed it to set seed. The use that bees made of pollen took a little longer to find.

4. Tempting bees. May 2019.

The discovery of insect pollination in the 18th century left its observers marvelling at a Providence which had supplied creatures designed to carry pollen from plant to plant. Some years later, Darwin and others would make clear the advantage to the flower: that without the uncertainty and the volume of pollen needed to use the wind, it could get genetic material to and from another individual, and so avoid the need to fertilise itself. Self-fertilisation for generation after generation, Darwin suggested, was something that all species sought to avoid, and for many flowering plants, insects bearing pollen were the answer.

Some insects visit flowers for nectar, and carry pollen only by accident. But honey bees are amongst those which gather pollen deliberately, and pollen for plants and bees is a kind of commodity, expensive to make and important to find. It may sometimes control the most basic aspects of bee ecology: the northern limit for the survival of hazel and other trees which provide spring pollen might be one of the main things which determine how far north honey bees themselves can live. This need for pollen was discovered at around the same time as pollination itself, but it took a while to discover how the colony used it, and at first it was mistakenly linked with beeswax.

Bees use wax to make cells in which to raise their young and to store the honey which allows them to keep large colonies alive across the winter. They also keep pollen in wax cells, but at first the relationship was thought to work in the other direction as well, so that pollen was used to make the wax. This was itself a development from an earlier belief, that we can date at least as far as Aristotle, that bees gathered the wax from which honeycomb is made directly from the plants they visited. In his *Discourse Or Historie of Bees* (1637) Richard Remnant still held to the belief that wax is collected from plants, but when he wrote that 'They gather wax all the yeere, from the first gathering to the last, from the willow to the blowing Ivie ... ' we can substitute 'pollen' for 'wax' and have a fair summary of the pollen season from spring to autumn. In the same way, his observation that 'the wax is gathered of the flowers or bloomes, with the fangs of the Bee, and so she puts it to her thighs, and rubs one against the other to fasten it on: and then carries it home, and make the combs in their Hives' describes how bees harvest and carry not wax but pollen. So once microscopes had revealed more about the anatomy of flowers, and that bees were gathering pollen from them, it was a short step to thinking that pollen, though clearly not itself wax, was the raw material from which wax was made. When Arthur Dobbs wrote his letter to the Royal Society in 1750, he described how he 'frequently follow'd a Bee loading the *Farina,* Bee-Bread or crude Wax, upon its Legs.' ('Farina' is a word used as often as 'pollen' at this time. Both come from

Latin words for flour or mill dust which were transferred to this new form of tiny grain. We probably favour 'pollen' today because that was the form chosen by Linnaeus). The French entomologist de Réaumur argued at around the same time that this pollen was transformed into wax through some process of digestion, and the idea persisted. In the mid-19th century, Robert Huish could still argue in his *Bees: Their Natural History and General Management* that bees took pollen into their stomachs, and there transformed it into wax. He was aware of another opinion, but since that was supported by Francis Huber, the target of ridicule and abuse through much of his book, he had no time for that. But Huber, as usual, was right.

Huber was born in Geneva in 1750. He lost his sight as a young man, but nevertheless carried out a lifetime of experiments on bees with the help of his wife and son, and of an assistant, Francis Burnens, in whom Huber placed his 'entire confidence, feeling sure to see well when seeing through his eyes.' The observations of this little group were so important that one modern writer has suggested that the whole history of beekeeping can be divided into pre- and post-Huber, and amongst their discoveries was the significance of pollen for bees.

Huber came to doubt that pollen was the raw material for wax. Others shared his view, notably John Hunter in England. They noted that when bees set up a new colony, and need wax to create new comb, you would expect them to collect pollen, and they do not. On the other hand, when a colony is well-established and the need for new wax is less, pollen *is* collected. So Huber began to suspect that pollen and wax were unconnected, and tried some experiments. He fed bees on sugar syrup and no pollen, and found that they could still produce wax, whilst bees on the opposite diet could not. So pollen is not the raw stuff of wax, and its real function, it turns out, is to feed the young. Hunter suggested this in a paper read to the Royal Society of London in 1792. Having described the farina, or pollen, brought into the hive, he commented:

> That it is the food of the maggot is proved by examining the animal's stomach; for when we kill a maggot full grown, we find its stomach full of a similar substance, only softer, as if mixed with a fluid, but we never find honey in the stomach; therefore we are to suppose it is collected as food for the maggot, as much as honey is for the old bee.

'John Hunter is not an authority to be consulted on the physiology of the bee', sniffed Huish, with his usual treatment of opposing views, but Huber investigated more carefully. Bees kept with syrup but unable to access pollen, he found, could not feed their young, and the brood starved. Larvae in the care of bees with a store of pollen, on the other hand, developed normally, and Burnens was able to describe to his employer how the adults 'took the pollen, grain by grain, in their mandibles and conveyed it to their mouths:

those that had eaten of it the most greedily climbed upon the combs before the rest, stopped upon the cells containing the young worms, inserted their heads in them and remained there for a certain time.' Like birds bringing food to the nest, they were feeding the larvae with pollen. It is not the raw material of wax at all. Huber was right about that too: bees create wax by secreting scales of it from abdominal glands, and the main fuel used to make it is not pollen, but honey. Pollen is only collected for food.

At this point in our year, the beginning of May, the colony is well underway with its pollen harvest. The bees that have survived last winter have begun to feed it to the young that will help to found a new colony, if this one swarms, or to raise further generations that will over the summer gather next winter's stores. These will be mostly honey, but some of the cells in the honeycomb will contain stored pollen, and this will then be used in the following spring until pollen is being produced outside, and the weather lets the bees collect it again. This was the cycle that Huber and Hunter had found, and the winter stores are a vital element. It was shown in the 1930s that there is a connection between the winter pollen stores and the success of the colony in the spring. But these stores are limited. Several hundred colonies in Scotland were monitored for seven years during the late 1940s and early 50s, and one of the results reported was that colonies usually maintain about a kilogram of stored pollen in the hive. Over the winter, though, this fell to about 75g, only enough when spring comes to raise about a thousand bees. At this time of year, with stores teetering on the brink of exhaustion, many beekeepers feed their colonies sugar paste enriched with pollen. Without this help, the tree pollen that we have seen so far is the main fuel for boosting the population of bees.

Pollen from cherries and other fruit trees dominate the early part of Synge's Spring stage, the second of the sections into which she divided the pollen year. As we move into May, neatly enough, the May-blossom of hawthorns takes over, and the end of the stage can be roughly defined by when the hawthorn flowering ends. Fittingly, for a tree that flowers on woodland edges, but is most common as a hedgerow and roadside plant, it marks the transition also from forest flowers to pollen that will mainly be collected from the open habitats of gardens and fields. Meanwhile, there are still a few other trees to contribute to the economy of the hive.

May 2019, Week 1

Predominant	*Crataegus monogyna*	Hawthorn
Secondary		
Important Minor		
Minor	*Taraxacum* type	
	Aesculus hippocastaneum	Horse chestnut
	Fagus sylvatica	Beech
	Brassica	

One of these is sycamore, another component in the mixture of tree pollen that drifts on the wind at this time of year. It is most common in May samples, but in our year, 2019, it was collected for the first time at the end of April. Around the apiary then and on into May, its long, hanging flowers are still opening, but there must be other trees, lower down, or more in the sun, where pollen is already being released, because some of the green or khaki loads that come in now consist of their neat, round grains.

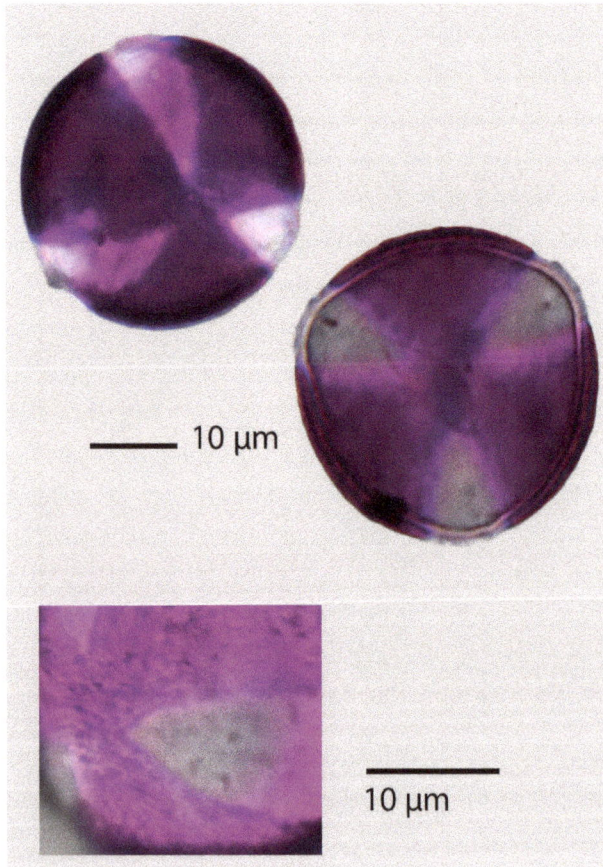

Figure 31: *Acer pseudoplatanus.* Family: Sapindaceae (maples)

The three openings of sycamore pollen grains are simple furrows, with no equatorial pores, and so they can be separated from cherry pollen, where each furrow contains a pore. These pores bulge out, and exaggerate the three corners, so that cherry grains are more triangular than round, and somehow less tidy than the contained shapes of sycamore.

Sycamore grains usually have a striate surface, with a pattern of light and dark lines, but some types of cherry and plum are like that too. The colours also overlap, so that green or khaki pollen can be either of the two. It can also be the pollen of a brassica, one of the plants in the mustard family that includes cabbage and kale. Many kinds are grown, but at this time of year, over much of Britain, the particular brassica that the bees have found is probably *Brassica napus,* or oilseed rape, where the pollen is more of a dull yellow. Bright yellow fields of rape are a familiar part of the landscape now, and bees collect the pollen wherever the crop is grown, but this was not always so. When Mary Percival collected her pollen records in Cardiff in 1945, charlock *(Sinapis arvensis)* and wild radish *(Raphanus raphanistrum)* were the only major sources of brassica pollen, and oilseed rape not found at all. A year or so later, Ann Synge's brassica samples in Hertfordshire included charlock again, and white mustard, another related plant. Honey surveys show the same effect. When Deans surveyed British beekeepers in the 1950s, brassicas were in the top five nectar sources, but these were crops of mustard and related plants. Then came the sudden change, so that the next major survey in the 1980s found that there was now so much rape grown that its honey had become the most common type produced. Oilseed rape had by then been grown for hundreds of years, but the oil from its seeds tasted too bitter for human taste. It was fit only to oil machinery, and make cattle food. Then, in 1973, Canadian plant breeders developed varieties that had so little of the bitter ingredient that the oil became palatable. European Union subsidies in the 1980s made the crop economically attractive too, and suddenly we had yellow fields and the bees had a new source of food.

Not everyone was enthusiastic. My grandfather gave up beekeeping at the time, partly because of age, but also because, almost overnight, he found himself surrounded by rape fields and didn't like the honey that it made. He found the taste bland, and was annoyed by its property of crystallising so quickly that it would often set solid before he could extract it from the hive. But other beekeepers adapted, and learned to extract early in the year if their bees had been foraging on the crop. The bees are certainly keen, which is just as well, because in some years it was grown on more than half a million hectares across the UK, out of the just over 6 million hectares that are used for crops. Planting across Europe received a further boost after 2008 when EU targets for renewable fuel made oilseed profitable for biodiesel. One effect of this is that rape fields have often replaced cereals: an insect-pollinated crop for a wind-pollinated one, and so increased the importance of insect pollination in agriculture. In Canada, where the oilseed revolution began, it has been shown that three beehives per hectare of rape can markedly increase the crop of seeds. As with the almond orchards, farmers might find it worthwhile to pay beekeepers to lend them some bees. Perhaps we will see more hives moved into fields at this time of year, although the smaller scale of our landscape may mean that a crop is always close enough anyway to an apiary to attract all the bees it needs.

The pollen grains of oilseed rape are small and round, with a net pattern on the surface. The image on the lower left shows the distinctively broad outer coat, with its well-marked columns. These components of the exine are often invisible with ordinary equipment, but are important features when an electron microscope is used for more detailed work.

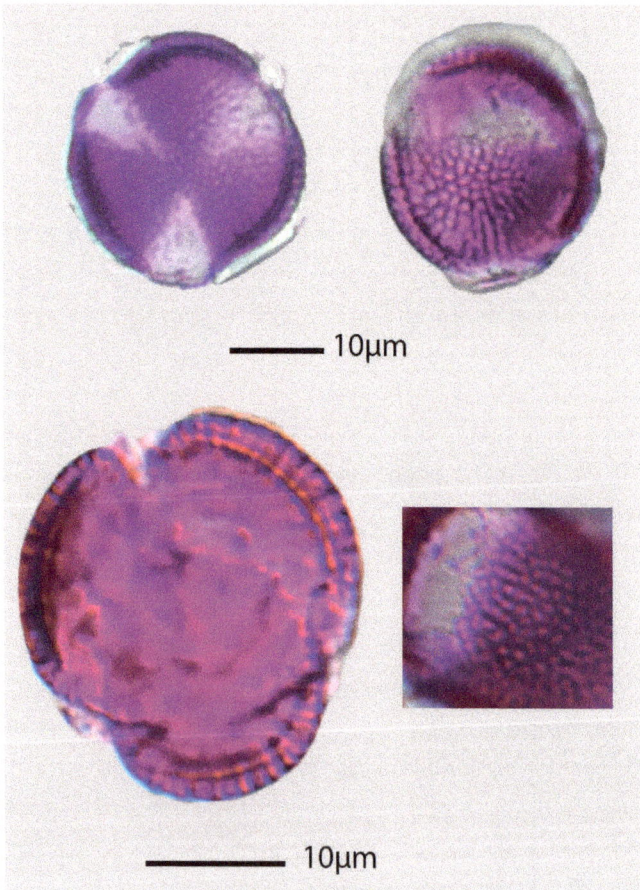

Figure 32: *Brassica*. Family: Brassicaceae (mustards)

May 2019, Week 2

Predominant	Crataegus monogyna	Hawthorn
Secondary	*Brassica*	
Important Minor	*Aesculus hippocastaneum* *Acer pseudoplatanus*	Horse chestnut Sycamore
Minor	*Plantago* *Clematis montana* *Pyracantha*	Plantain

Gradually, as the fruit trees stop flowering, sycamore and hawthorn replace them as sources of pollen. The landscape of my grandfather's bees changed dramatically during his lifetime, with fewer hedges, and crops like rape grown in larger fields. Without hedges, there would have been fewer sources of the hawthorn pollen that marked the later part of Synge's Spring stage. Around the apiary here, though, farms that have survived industrialisation and the spread of housing are mostly laid out much as they were in the mid-19th century, when the tithe maps set out the landscape for us field by field. Most of the fields that still exist are small, and still separated by hedges in which hawthorn is the most common tree. This is another plant in the Rosaceae family, and its pollen has the same general shape as that of the fruit trees earlier in the year. At this time of year, though, its large size – a little more than 40 micrometres across, and white-green colour are distinctive, and it can often make up 70 or 80% of the samples in the trap.

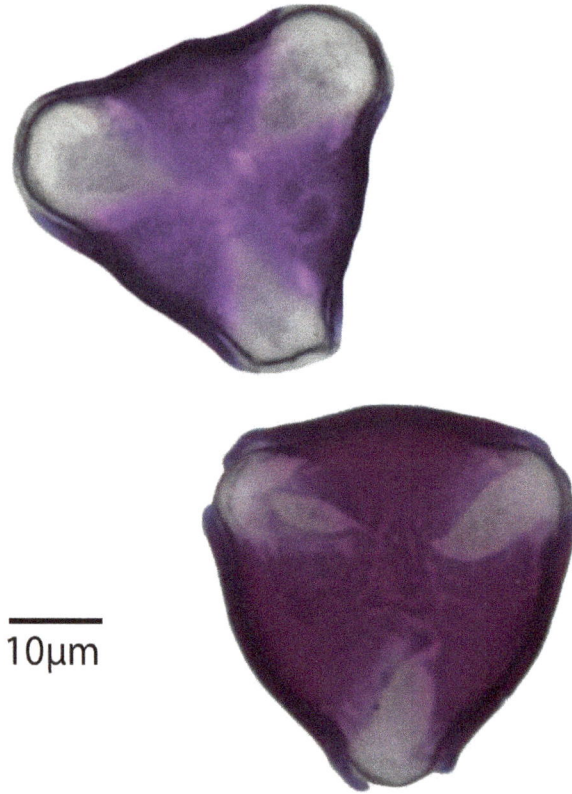

Figure 33: *Crataegus monogyna.* Family: Rosaceae

Hawthorn is such a constant feature of our landscape, flowering each year along road verges as well as field boundaries, that it comes as a surprise to find it unreliable as a nectar source. But beekeepers have known for a long time that some years will yield a good crop, and others very little, and that within a single year there are dramatic differences from place to place. Occasionally there is even enough to collect hawthorn honey from the hive, and R.O.B Manley remembered an almost pure stock from 1911 as being the only honey he had ever really enjoyed in a lifetime of keeping bees. But this kind of yield, he wrote, was rare, and pollen is the more constant reward for bees which visit this plant. They will collect it as their main type now over several weeks.

Even after the hawthorn fades, hedges will be of some use to bees, through the climbing plants that scramble over them. The dog rose will flower soon, and *Clematis vitalba,* Old man's beard, later in the year. Bees use garden clematis too, occasionally bringing back pollen from some of the large-flowered types, and more often from *Clematis montana,* which is vigorous enough to form sheets of flowers and attract bees to collect the white pollen loads. The grains are small, and have the same three stippled furrows as their relatives in the buttercup family, the celandines and wood anemones that we saw earlier in the year.

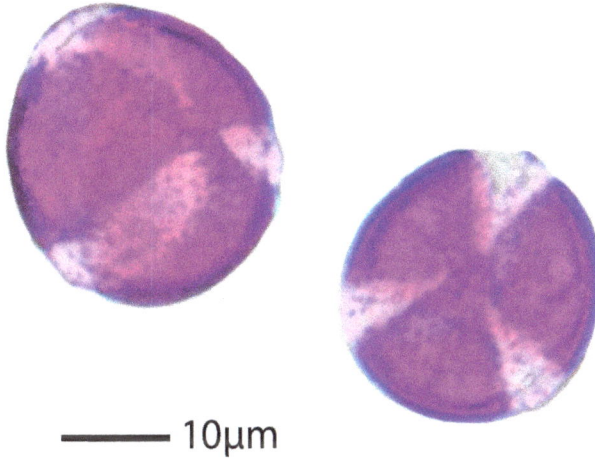

Figure 34: *Clematis montana.* Family: Ranunculaceae (buttercups)

May 2019, Week 3

Predominant	*Crataegus monogyna*	Hawthorn
Secondary		
Important Minor	*Acer pseudoplatanus* *Brassica*	Sycamore
Minor	*Aesculus hippocastaneum* *Plantago* *Clematis montana*	Horse chestnut Plantain

Also in May, but vivid amongst the dull shades of hawthorn and clematis, the brick-red colour of horse chestnut pollen is unlike anything else that bees collect, and recognisable even at the entrance of the hive. It hardly needs examination under the microscope, but there it has a neat oval shape and obvious stained dots along the furrows, and in the centre of each furrow a circular pore:

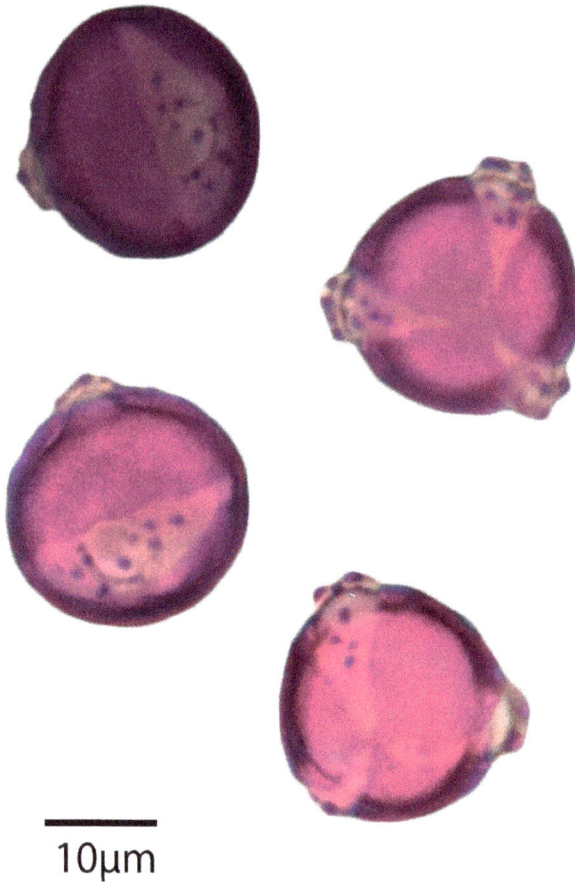

10µm

Figure 35: *Aesculus hippocastaneum.* Family: Sapindaceae

If you imagine horse chestnut grains as a collection of eggs, or rugby balls, scattered on the ground, most will end up lying on their sides, rather than standing up on one of the ends. But these grains have flattened poles and are mounted in jelly, and some balance in just the right way for us to see them on end, as in the two on the right, with the furrows almost meeting at each pole. By focusing on the equator of a grain which is standing up like this, on the lower right, we can see a different view of the granules that are scattered along each furrow. Horse chestnut is in the same family as sycamore, but the pollen could hardly differ more.

The tall flowering spikes are as striking as the pollen, and *Aesculus* is widely planted as an ornamental tree. Studies which analyse the pollen grains present in the air have shown that some horse chestnut pollen is carried by the wind, but as the large, showy flowers suggest, insect pollination is the more important route. The flowers even use colour changes to communicate with their visitors. Young flowers have yellow streaks on the petals that lead the insect to the nectar or pollen inside the flower, and these turn red as the flower ages. Kugler showed in 1936 that flowers with the yellow guides are visited far more often by bees, and this means that the plant is directing the insect to the younger flowers which have not been pollinated yet. It is suggested sometimes that the bees ignore the red guides because their eyes are not sensitive to red: the idea is appealing, because it creates the image of the flowers turning themselves on and then off, like lights. But lungwort *(Pulmonaria)*, which is also pollinated by bumblebees and some other relatives of the honey bee, is another flower that changes colour as it ages, and in this case the younger flowers, which are visited most often, are red. Whatever the details of the colour signal, both horse chestnut and lungwort offer greater food rewards with the flowers that they want the insects to visit, and it seems likely that the bees can see the colours well enough to associate them with the food for which they have come.

In this case, the food is not usually pollen. There are plenty of trees around the apiary, but only one or two horse chestnut pollen loads turn up in the trap each day. At this time of year, the bees are using two distinct types of tree. Beech and oak are among the last examples in the year of trees that may be visited by bees, but use the wind to carry their pollen, while the horse chestnut is a tree with bright and obvious flowers to which insects are drawn. Tempting bees to your pollen is a risk, if they end up eating it all instead of taking it to another plant, so from the chestnut's point of view, the fact that so little pollen ends up in the hive is a good thing. Instead, when the bees carry its pollen, they are carrying it as the flower would want – from tree to tree, and not back to feed their young. Mary Percival recorded the same thing in her Cardiff hive, seventy years ago, and noted that, although little pollen came back to the hive, horse chestnut was still important to the bees. The tree was valuable not for its pollen, but for nectar, one of the other rewards that flowers can provide. If hazel and elm and the other first trees of the year, flowering before most insects appear, are plants for which bees are usually nothing more than predators, the second stage of the pollen year is an annual reminder of how other flowers are connected with insects in a more positive way. The two groups have evolved together in a relationship partly responsible for the success of both.

Plants have probably lived with insects, or insect-like creatures, from the time they first colonised the land. Many of these insects were herbivores, and how plants resisted them, and the earth remained a green planet, is a question over which ecologists have argued for at least sixty years. Pollen would have been part of this insect diet: some of the springtails which still survive today, and resemble the earliest kinds of insects, feed on it. A recent study has suggested that springtails may also transfer the sperm of moss. Mosses evolved long before flowering plants; their sperm is not enclosed inside pollen grains and is usually transferred through water, which is one of the reasons why we associate moss with damp places. If springtails are another, more active, way of moving sperm, then we have a glimpse across hundreds of millions of years to when flowering plants appeared in a world which already contained the animals that could transport the male cells in their pollen. But the relationship that is so familiar now took time to develop, and the earliest types of land plant would have relied on the wind to transfer their spores. This would also have been true of the first gymnosperms – ancestors of today's conifers, and of the yews whose pollen we saw earlier in the year. The older insect families, including beetles, would have visited them at first only as predators, another of the hazards faced by plants on land. Fossil insects, dated to about 270 million years ago, have been found with pollen in their gut. But gradually the relationship evolved into one where the plants gained the benefits of pollen transfer alongside the cost of insects eating it. The angiosperms, appearing later, were probably from the beginning pollinated by some of their insect predators. Wind pollination in some of these most recent kinds of flowering plants then evolved later, perhaps in places or seasons where insect pollinators were lacking, just as we have seen so far with trees that flower early in the northern year. But before that, as the angiosperms became the dominant form of land plant, most relied on insects to carry their pollen. That some of the pollen was eaten, or carried back to a nest to feed insect young, was a necessary evil – the price paid for the pollination service. There are many plants which still survive in this way: the bees this year have so far visited wood anemones and gorse, which provide no nectar, but plenty of pollen. The pollen itself, or the anthers in the case of the anemone, has a colour and perhaps a scent that advertises its presence. The bees which come will take much of it back to the hive, which is of no help to the plant, but as they gather it, enough may be transferred to fertilise the plant. The tendency of pollen-collecting honey bees to keep to the same kind of plant makes this all the more likely. If wind pollination is like boarding the first bus and hoping for the best, then insects are taxis – expensive, but more likely to take you to the door.

The expense might explain why some plants, like the horse chestnut, developed other rewards for their pollinators – currencies other than pollen that might be cheaper to produce, or more attractive for particular kinds of insects which would then become faithful couriers, drawn to the same type of plant again and again. Some plants provide oils, scented or nutritious. Others offer resin which insects use to make nests, and some, like the figs, provide a space in which to breed. But nectar is the most common reward: about 2½ milligrams per flower in the case of horse chestnut, of a liquid which is something like 40% sugar. The sugar fuels flight, allowing insects to find food and each other. In the case of honey bees, it even allows the accumulation of honey stores that can maintain thousands of bees across the winter.

By fossil standards, these alternative rewards are quick to evolve. No more than ten million years after the first fossils of insects with structures that suggest a connection with flowering plants, there are plants with floral nectaries – places where nectar is provided, just as in horse chestnut flowers today. Floral nectaries not very different from those of the horse chestnut itself appear about 90 million years ago. And now the plant can have a new strategy altogether for dealing with its pollinators. If pollen is no longer needed to attract the insect, then the plant can make less of it. In fact, it can try to stop the insect from eating or collecting pollen at all. Then all the plant has to do is make enough pollen to be carried to other plants, and pay the courier with nectar instead. Long before humans discovered the same trick, plants had found that sugary liquid could be very profitable indeed.

Horse chestnut flowers show how this works. They have the classic 'bee-flower' structure, where the flower forms two parts – an upper roof, and a lower floor. This is quite different from the open, circular arrangement of the anemone, and its significance was noted by Kölreuter and Sprengel long ago. The bee lands on the lower part and can move inside. But the anthers are on the upper part, and brush pollen on to the bee's back. It is difficult for the bee to get at the anthers to collect pollen in the usual way, or to reach the pollen left on its back. Easier to just drink the nectar and move on to the next flower. And there, without the bee having taken any interest in pollen, the grains still on its back can brush on to a stigma, and fertilise the plant.

So if the chestnut offers nectar as its bribe, and pollen just as a light dusting on the back, why do we see any of its pollen in the hive at all ? Where Sprengel saw the hand of the Creator in flower design and the role of insects, modern biologists often refer to these kinds of relationships as 'arms races'. The partners gain from cooperation, but they are never averse to cheating one another if the opportunity comes. Sometimes the chestnut just loses out, and a red pollen load turns up in the trap.

Hawthorn is different: a genuine dual source of food. Its pollen dominates the samples for most of May, but we have seen that in some years at least, it also provides so much nectar that the bees can make a honey crop from it. Sycamore is another tree which provides both.

May 2019, Week 4

Predominant	Crataegus monogyna	Hawthorn
Secondary		
Important Minor	Acer pseudoplatanus Clematis montana	Sycamore
Minor	Aesculus hippocastaneum Sambucus nigra	Horse chestnut Elder

These two offer a mixture of rewards, but some provide only pollen and some just nectar. Then there are those which specialise still further, and only make certain kinds of visitors welcome. Elder is an example. Although elder flowers are abundant and obvious at this time of year, they seem mainly to attract flies, and bees collect only a little of their pollen, in yellow loads.

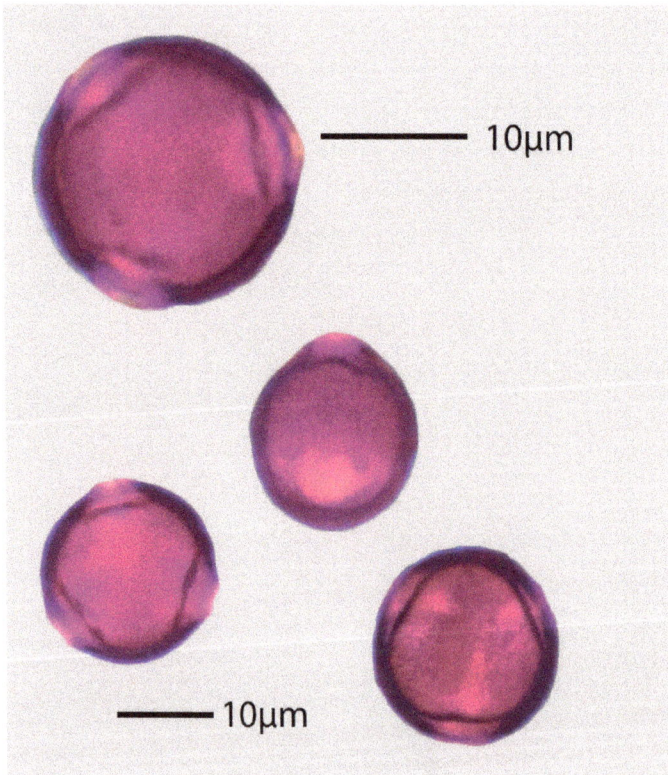

Figure 36: *Sambucus nigra*. Family: Adoxaceae (Moschatel)

Elder draws flies, and horse chestnut we call a typical 'bee flower', with structures that seem to have evolved to suit bees. Other kinds of plant keep their nectar at the end of long tubes, out of reach for honey bees and only accessible to moths and butterflies, or humming birds in warmer parts of the world. Lilacs are a familiar example here, and

Howes noted that unless nectar production was particularly abundant and the level in the flower tube unusually high, honey bees couldn't collect it. Flowers of this kind of shape provide a vivid example of how co-operation can give way to cheating. Sometimes insects that can't reach the nectar have learned to bite into the base of the tube and steal the nectar without passing the pollen apparatus at all. In that particular arms race, the insects have moved ahead.

Colour and scent may also be tailored to the intended visitor, like white flowers and evening scents that are typical of plants pollinated by moths, or smells of decay that are repulsive to us, but attract certain flies. As more and more information was collected about the connections between plants and the pollinators they tempted, these 'moth traits' or 'bee traits' were used to organise it. In the 1870s an Italian, Federico Delpino, was the first to try to set out a general arrangement of plant types using these traits. Regardless of whether a plant was in the lily family, or the rose, whether it grew in Asia or Brazil, if it was to be pollinated by the wind, then the system proposed that it would have simple, unshowy flowers dangling in the air. If, on the other hand, it was pollinated by a night flying insect, it would tend to be white and release any scent after dark. To Delpino, these sets of traits were established by the Creator, as if the work had been shared out amongst the pollinators by assigning certain plants to each. There was criticism even at the time that he was too keen to force flowers into his categories and ignore the evidence that many were pollinated by more than one kind of insect. His pollination syndromes, though, as we now call the sets of characteristics that match a plant to its pollinator, have been studied ever since. But now, instead of assigning each plant to a particular agent, we tend to frame the argument in terms of whether a plant is a specialist or a generalist with regard to how many different kinds of pollinators it receives. The data is hard to collect. In the 1920s, the American entomologist Charles Robertson published *Flowers and Insects,* a book which listed the insects visiting 296 species of flowers within ten miles of Carlinville, Illinois. He recorded 15,172 visits, and identified the insect in each case. His results are still used today, partly because nothing quite so heroic has been done since. On average, each of his flower types was visited by more than thirty kinds of insect, but a few by only one. In the same way, some insects are specialists, concentrating on only one or two kinds of flowers, and others generalise. The honey bee could only ever generalise: thousands of bees in a colony could never be maintained by a single plant, and the honey bee takes its pollen and nectar wherever it can. An 'unusual super-generalist' is how one entomologist has described it.

Arguments about the evolution of specialist and generalist flowers continue. Sometimes links can be made to the environment in which a plant is found. In a landscape with a large number of plant species, for example, each type of plant may grow as individuals widely scattered from each other. In this case, a specialist delivery system might be the only answer. At the other extreme, a tree which forms a forest mostly of its own kind can let its pollen travel with any carrier, even the most indiscriminate of all – the wind. In that kind of forest, the pollen has a good chance of blowing to the right place. Or you might look at the life history of the plant itself. An annual weed has only a single season to transfer its pollen. Perhaps it should use as many pollinators as it can. A long-lived species, though, may have several chances, and can get the most efficient results by using less pollen and a specialist carrier.

Of course, being a specialist has its own kind of vulnerability. What if something happens to the single kind of insect able to carry your pollen ? Or how could you grow in new places if the insect wasn't there ? The second point could matter to us, because we have moved plants all over the world, as crops or for pleasure. And yet, with the exception of the bumblebees taken to New Zealand to pollinate clover, we rarely think of moving pollinators too. It seems that most plants have remained accessible to several insects, and that wherever they are taken, there will usually be a generalist pollinator that can deal with them there.

For a few weeks as we move from May into June, the California lilacs, *Ceanothus,* are a good example of this. The name is a partial guide - some species grow in other parts of north and central America, but most are found in California. The plant hunter William Lobb, best known for providing the Victorians with the giant sequoia and the monkey-puzzle tree, came across it there and sent the first samples to the Cornish nursery which employed him. Half a world away from its home, its pollen is still accessible to a range of bees, including the honey bee, which now starts to bring back a steady supply of its small, whitish grains.

June 2019, Week 1

Predominant		
Secondary	*Meconopsis cambrica* *Chenopodium* *Taraxacum* type	Welsh poppy
Important Minor	*Dianthus barbatus* *Ceanothus* *Papaver rhoeas* *Sambucus nigra* *Trifolium*	Sweet William California lilac Common poppy Elder Clover
Minor	*Cytisus scoparius* *Rhododendron* *Heracleum spondylium* *Plantago* *Geranium*	Broom Hogweed Plantain Cranesbill

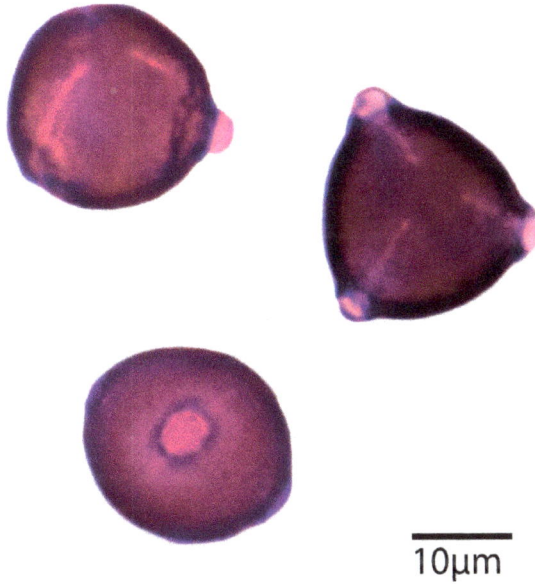

10µm

Figure 37: *Ceanothus*. Family: Rhamnaceae (buckthorns)

Another plant used in small amounts during May is represented by light yellow pollen loads. These come from the plantains, mostly the ribwort plantain, *Plantago lanceolata,* that is flowering on any patches of open ground that aren't mown too often. It is one of the less usual forms of pollen, with pores not restricted to the equator, but across the whole surface of the grain.

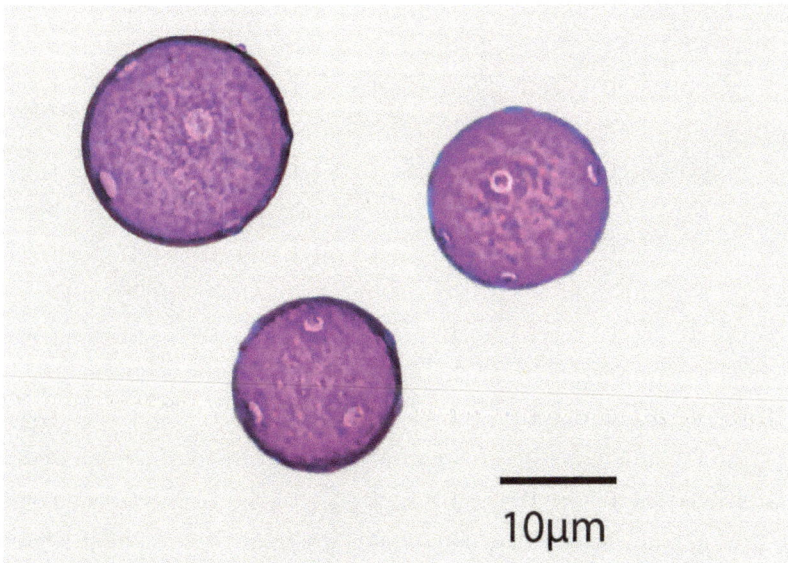

10µm

Figure 38: *Plantago lanceolata.* Family: Plantaginaceae (plantains).

Another grain with the same type of structure also appears now, collected from the sweet Williams in the garden, *Dianthus barbatus.* They are larger, and the pollen loads are black, but the general appearance of these unrelated types is similar. On the larger *Dianthus* grains, we can see now how the granules in the centre of the pores of both types of plant seem to stand away from the surface and form a cap above the opening. This is called the operculum.

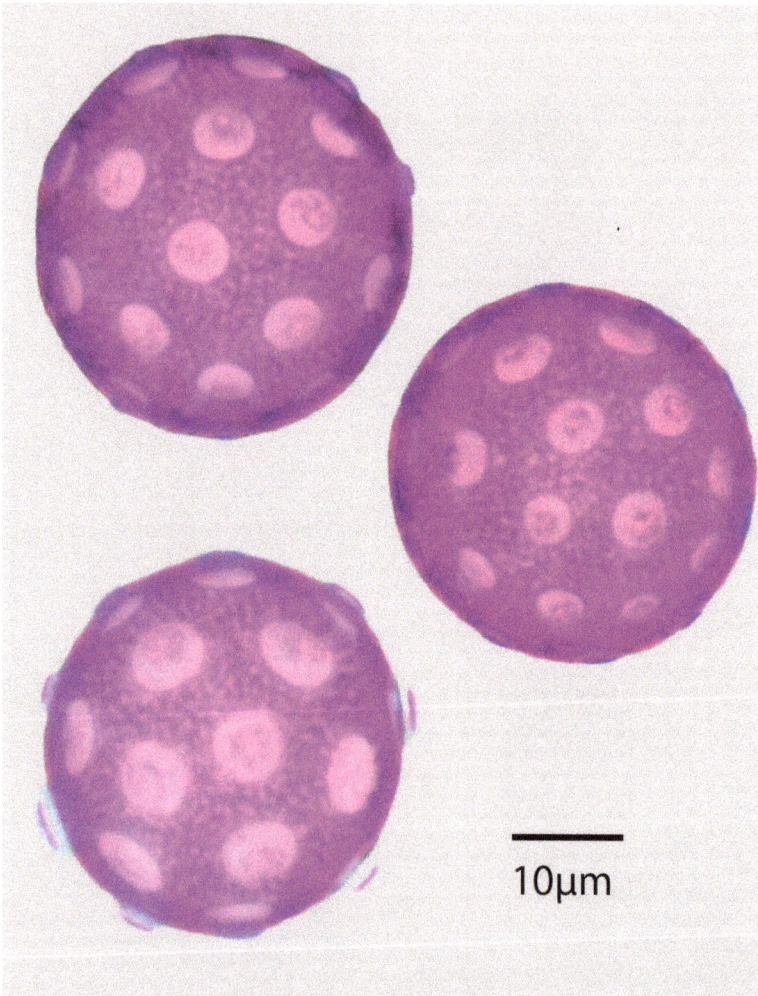

Figure 39: *Dianthus barbatus.* Family: Caryophyllaceae (carnations)

Plantain pollen is collected here every year, but the *Dianthus* loads only rarely. Another unusual type in this latest sample is also black, and also comes from the garden: pollen from the cranesbills, *Geranium.* These are distinctively large grains, about 75µm across. They have three openings, which in most types of pollen would mean three obvious furrows, but in this case the apertures are so short that they look more like pores.

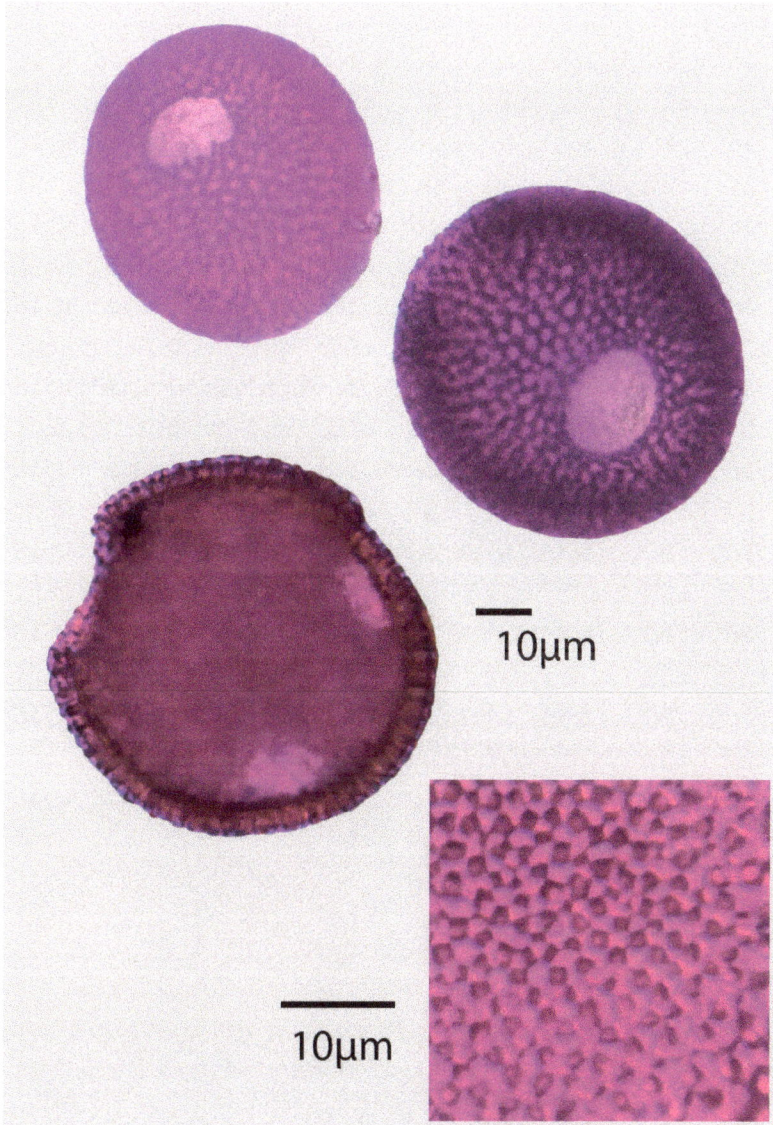

Figure 40: *Geranium.* Family: Geraniaceae

Plantago is a plant of open habitats: its pollen, preserved in waterlogged places, was another of those important to the palynologists who began in the 1920s and 30s to use pollen grains to reconstruct the history of vegetation in Britain. Its distinctive grains are a record of both the time before forest spread across most of Britain after the last ice age, and then the clearance of that forest later when the first farmers settled here. Open ground and large populations are some of the factors that make wind pollination efficient, and the plantains are examples of flowers which have gone back to using the wind. That this means holding their pollen out in the air, where bees and other insects can take it, is a disadvantage that they must live with if the animals then use it as food rather than leaving it with another plant. Broom, meanwhile, represented by orange loads in the latest sample, uses the normal strategy of flowering plants. Insects are attracted by its bright flowers and strong scent: some pollen is used by the visitors and hopefully – from the plant's point of view -some transferred. This is the 'pollen dilemma': if plants use pollen both to feed insects and to be carried by them, then how do they keep the cost of the feeding down, and the chance of transfer high ? As we have seen, some limit their visitors by specialising: a strategy with risks of its own. Broom flowers have the typical shape of the pea family to which they belong: any pollen or nectar is hidden inside an enclosure formed by the two keel petals. Only an insect large enough to land on the keel and force the petals apart can find its way in. When it does so, some of the anthers expand into the opening and dust the bee with pollen. With so many grains caught in its hairs, some in places difficult to groom, the bee is unlikely to gather them all into a pollen load and some will go to the next plant it visits. Broom is an invasive species in many parts of the world, and part of its success is probably that one of the insects able to use its flowers is the similarly widespread honey bee. Broom has gambled on the right partner. It may also reduce pollen costs by tempting insects with nectar instead: many of the garden varieties of broom provide this too. Even so, the dark orange loads in the pollen sample, destined to feed larvae rather than fertilise eggs, show that broom must use at least some of its pollen to pay the bees.

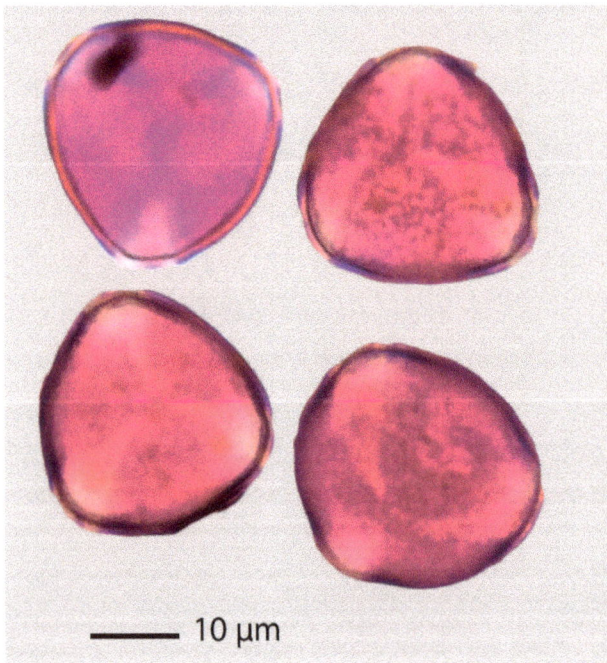

Figure 41: *Cytisus scoparius.* Family: Fabaceae (peas, beans)

5. Gathering pollen. June 2019.

The yellow or Welsh poppy, *Meconopsis cambrica,* is a common plant near the apiary, as easily spread as any weed but attractive enough to be encouraged in gardens. For a week or so at this time of year its pollen can be the most important type collected here, and bees move from flower to flower, gathering grains into light orange loads.

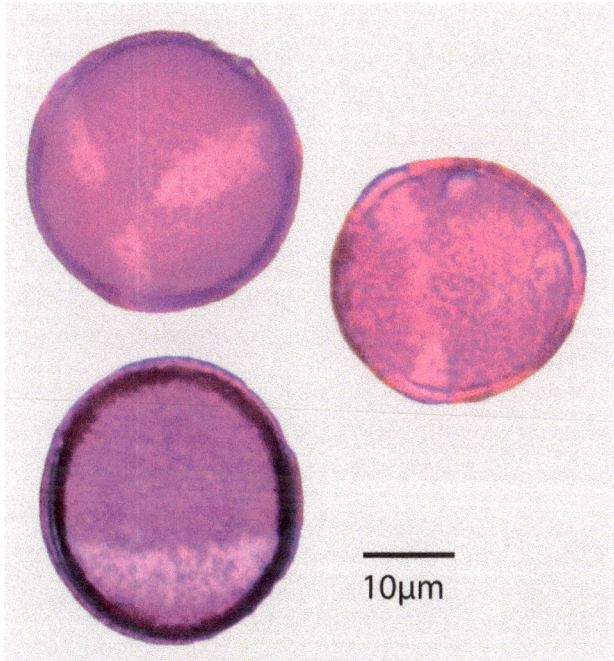

Figure 42: *Meconopsis cambrica.* Family: Papaveraceae (poppies)

On an average trip, a honey bee foraging for pollen may be out of the hive for thirty minutes, although of course this varies enormously depending on what is flowering and how far away it is. For my bees, for example, there are poppies in the garden, but the nearest field of rape is more than a mile away. In some ways, pollen foragers have things easier than their nestmates collecting nectar: they can collect a full load from fewer flowers than the hundreds or more that a nectar load requires, and in less time. But much depends on the factors we considered in the last chapter: the structure of the flower, and the balance between pollen and nectar rewards. With a good source like a poppy, its pollen laid out for collection, a single flower alone can provide several loads, but more often a bee will have to visit several plants before flying home. Later in the year, we will find that white clover is one of the most important crops of the year, supporting the colony over many weeks, and yet gathering enough pollen from it is particularly difficult. Each flower head contains many individual florets, and Percival had to open almost 400 of these to extract seven thousandths of a gram. She estimated that a bee would have to visit a hundred heads and poke inside 585 of the florets before it had a full load.

A forager packs thousands of grains into the ball on each hind leg, and even in plain sight on an open flower like a poppy, the collecting process is no more than a blur of moving limbs. It seems impossible, without slowed-down video, to work out what is going on. 'Probably the kinematograph,' suggested FWL Sladen in 1911, only fifteen years or so after the new technology had been demonstrated in France, 'will be able before long to reproduce the whole process of pollen-collecting at a speed slow enough to be followed by the human eye.' But remarkably, Sladen himself, a great authority on bumble bees, had already had the patience and ingenuity to work out most of what was going on. Then a year later, interested in the commercial value of honey bees for pollinating fruit, the US Department of Agriculture published a pamphlet by DB Casteel which also described how they gather pollen. The new cinema was of no help to Casteel either, who relied on observation and photographs, and was concerned that some details might have been missed. But although he and Sladen had each been unaware of the other's work, it turned out that they were largely in agreement, and largely correct.

The front legs and the mouthparts are the main tools for harvesting the pollen. Each leg has a collecting brush formed from dense hairs, and this combs the anthers for pollen grains. In addition, pollen which is dislodged by the movements of the bee and falls on to its head or the front of the thorax is groomed off by these brushes. The antennae have their own cleaning site on the front leg: they are drawn through a notch at the base of the tibia where a comb of bristles brushes them clean.

Figure 43: front leg with cleaning notch

The front legs also make frequent contact with the mouthparts: there they collect moisture which helps to hold the pollen grains together as they are gathered into a ball. Casteel found that this moisture had a sugar content typical of honey rather than nectar, and concluded that it comes from honey brought from the hive, which is regurgitated by the bee and transferred to the legs. Sometimes plant nectar is used too, when bees collect nectar and pollen from the same plant, and nectar is then mixed with the pollen as the bee gathers it, with the same effect of gluing the grains to each other. Finally the front legs transfer the moistened pollen back to the middle legs. Collecting, grooming, moistening, passing back – and by this last stage the bee is already in the air again, because with its legs free as it hovers, it can rub them against one another and pass the pollen back.

The middle legs also have a brush of hairs, and a long spur which may be used when the pollen ball is eventually unloaded in the hive.

Figure 44: middle leg brush and spur

As well as receiving pollen from the front legs and passing it on to the back, the middle legs groom the hind part of the thorax where the front pair can't reach. There is a grooming 'blind spot', however, and sometimes bees come to the hive with a stripe of pollen which marks where they cannot reach. Sometimes, even, they return with pollen loads but also pollen dusted over their bodies, as if the grooming process has been overwhelmed.

As the bee hovers and the pollen accumulates on the back legs, one movement is clear: a patting and smoothing motion of the middle legs across the pollen pellet. This must help the grains to cohere in the ball, and may also smooth the surface to reduce the air resistance of these clumsy loads. It looks as though the middle legs are placing the pollen directly on to the outside of the back legs, but this is not the case. Casteel thought that a little pollen would be transferred from the middle leg as it smoothed down the pollen ball, but described how most would actually be passed back to the inside of the hind legs, and only then used to form the load. This also takes place in the air, when the legs are free. First, the middle leg is rubbed between the two back legs. The inner surface of the back legs – the surface against which the middle leg rubs - has several rows of stiff bristles called the pollen combs, and they scrape off the pollen from the middle leg.

Figure 45: pollen combs

Each back leg now has pollen on its inner surface. The pollen loads, however, will be carried on the outer side of the leg, and the final task is to move the pollen there. With the bee still in flight above the flower, or moving on to the next, each hind leg moves its pollen to the outside face of the leg on the other side. This is where the pollen load forms and where it is carried when the bee flies back to the hive.

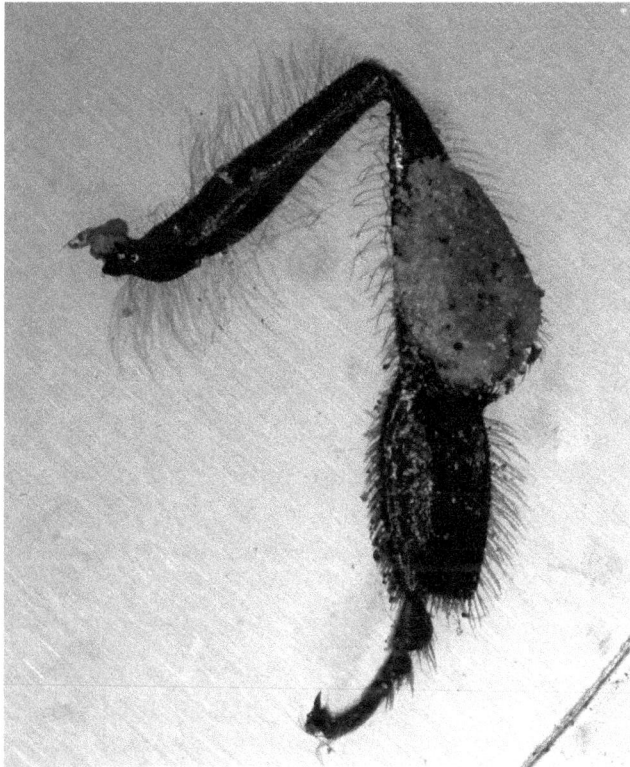
Figure 46: Pollen load forming on leg

Different kinds of bees have different ways of carrying pollen home. One of the ways in which honey bees and bumble bees are alike is that both have a depression on the outer side of the hind leg into which the pollen is packed. In English, the structure has been called a basket since the mid-eighteenth century, and the Latin term, corbicula, which was introduced a little later, means the same thing. It is a hollow region on the tibia of the hind leg, over which long hairs arch and help to hold the load in place.

Transfer from one hind leg to the other begins with another comb of bristles, this one called the pecten. The pecten is at the base of the tibia, and so on the bottom edge of the basket. The pollen comb of the left leg is rubbed against the pecten of the right leg, and the pecten of the left leg cleans the comb of the right. With some pollen transferred to the pecten, the bee then flexes its back leg, and this movement forces the pollen from the pecten up into the basket.

The bee descends again, gathers more pollen, hovers and grooms. The pollen makes its way to the pecten, and is pushed up against what is already in the basket. Gradually, filling from below, the pollen load accumulates. When the baskets are full, there can be ten milligrams or more of pollen on either side, a total load perhaps a fifth of the body weight, and this the bee carries back to the hive.

As far as weight is concerned, bees carrying nectar are often more heavily laden, but there is something cumbersome about these external pollen loads, and a bee steering itself and its burden into the entrance of the hive can seem a chancy affair. These sometimes awkward arrivals, with round balls on the legs, may lie behind the ancient Greek belief that bees took on small pebbles as ballast to keep them steady in the wind and prevent them from being blown past their homes. Some miss, and every colony has its strugglers in the vegetation around the base, climbing up or gathering their strength to fly again. Once inside, the pollen is stored in the cells of the honeycomb. Nectar carriers pass their loads on to the younger bees that have not started foraging yet and only work inside the hive, but bees with pollen take it inside themselves and find the cell where it will be kept. This will usually be in a particular zone of the hive. Honeycomb has distinct layers, where the cells are used for different things. The central cells, which are the easiest to keep warm inside a cluster of bees, have the eggs and the developing brood. Honey is kept in the outermost cells: one advantage of this is that honey is heavy, and best stored closest to the points where the comb is attached to a wall. Pollen is placed in cells between the honey and the brood, close to the larvae to which it will be fed. In the dark, through a crowd of thousands, the bee finds a suitable cell. She pokes her head in to confirm her choice, then climbs out and grips the next cell up with her front legs. This lets her put her abdomen on to the edge of the chosen cell, and with her middle legs brush the pollen loads off to fall inside. The long spur on the middle leg may play a part in this, perhaps helping to free the pollen ball from its basket. All of this is done even if the bee has lost her load, so that bees which have left their pollen in the pollen trap will still find a cell, turn around, and attempt to brush off the pollen which is no longer there.

The forager may then set out on another trip. Ribbands, in 1949, having chloroformed and marked a sample of bees, watched them at work in a garden planted with different kinds of flowers. He found that one bee, collecting pollen from California poppies, made 47 trips in a day. He considered this to be about the maximum that could be achieved, and results from an earlier study in Iowa reported an average of only eight. However, those bees were collecting from maize, which only provides pollen in the morning, and even there one bee managed to fit in 20 trips before the supply was cut off. If pollen is available throughout the day, as with clover, where florets of different age open at different times, a bee collecting pollen can make many more trips than when it is looking for nectar.

This is not to say that there are pollen specialists, working harder than lazier workers which exclusively search for nectar. Although honey bee workers have particular roles which change as they age – nurse bees, guards, even undertakers removing corpses from the hive - and may for a while spend most of their active time carrying out that particular task, once a given bee takes up her last role as a forager she will rarely spend all of those two or three weeks collecting only one kind of food. Some trips will be for pollen, others for nectar. On any single trip, however, she is likely to specialise. Ralph Parker made one of the first detailed studies of foraging behaviour in the 1920s, and found that only 17% of his bees were collecting pollen and nectar at the same time. The rest chose one or the other, with about twice as many gathering nectar as pollen. But this varies according to which kind of food is available, and what the colony needs, and although a particular bee may stick to either pollen or nectar for many consecutive trips, she will switch back and forth between them during her foraging life. Gathering pollen may look complicated, but all honey bees can do it.

Inside the hive, meanwhile, with the pollen deposited, one of the younger nest bees takes over. With her head, she pushes the pellets down into the bottom of the cell, and there breaks them up and combines them with any pollen which may already be inside. Flower constancy means that pollen loads on the bee usually consist only of one type of pollen, but different loads are dropped together in the cells, and a cell of stored pollen can in the end contain lots of different kinds. This is now the protein store for the colony: but protein mixed with about a quarter part sugar, which comes from honey which the nest bees mix with the pollen when they are packing the cell, and with which they cap the cell when it is full.

Without it, the bees would adopt desperate measures. They begin to break down the protein in their bodies. Young larvae are cannibalized and the protein fed to older larvae. Finally, brood rearing stops, and the colony will die. Under normal circumstances, though, enough pollen will be brought in to meet the colony's immediate needs and to keep a small amount as stores. Although larvae must have pollen to grow, it is not fed to them directly. Pollen is difficult to digest, and the nurse bees, between about five days and two weeks of age, first convert it into brood food, or brood jelly, and it is this on which larvae are reared for the first few days of their lives.

Pollen which is not needed at once can be kept in the cells and some, as we have seen, is stored over the winter. Beneath its cap of honey, the cell of food ferments, and becomes bee bread, a substance recently analysed as 66% protein, 26% sugar and small amounts of lipid and starch. Some believe that bee bread is modified by microorganisms from bee saliva to become more nutritious than pure pollen, perhaps because of an increase in vitamin content. The suggestion is that the bees are innoculating the pollen with their gut flora in order to bring about this conversion to bee bread. More recent studies have raised doubts. For example, a swarm can raise larvae on raw pollen before there is time to ferment it into bee bread. In this case, any honey added to the pollen comes from the insect's crop, where there are very few bacteria. It seems that the fermentation process may be more about preserving the pollen than changing its nutritional content. Like us, the bees need to keep their winter food stores from being spoiled by bacteria and fungi.

One way in which we do this is to store food under conditions where microorganisms can't grow. We keep green crops as silage, and some vegetables can be pickled. We can think of each cell of stored pollen as being like a miniature silage store or pickle jar. Strains of *Lactobacillus* thrive in the stored pollen, creating, as their name suggests, lactic acid as a metabolic by-product. This builds up and forms an environment so acidic that few yeasts and bacteria can grow and spoil the stored food.

It may even be the case that storage results in something less nutritious. *Lactobacillus* extracts energy from the stored pollen, and bee bread may actually contain fewer calories than are present in the raw materials added to the cell. Even if that is the case, it could still be that bee bread is more digestible than raw pollen, in the same way that cows can more easily digest silage than unmodified grass. This is not certain, though, and some studies have found that nurse bees prefer to feed on freshly-deposited or partly-fermented pollen rather than fully aged bee bread, like preferring fresh cabbage to sauerkraut. Perhaps bee bread is not an improvement on pollen at all, but something to put up with, when fresh food can't be found.

The bacteria involved are not all from the gut of the bee. Some are able to live freely on flowers, feeding on sugar-rich nectar, and bees transfer them to the stored pollen when they mix the pollen with nectar or honey. This means that the microflora of the bee gut is affected by the type of landscape in which it forages. The differences between landscapes can be assessed now with new methods of bacterial DNA sequencing, and the discovery that bees top-up their microflora in this way suggests that we may one day see the equivalent of probiotic drinks for bees, where the beekeeper provides what advertisers have taught us to call 'friendly bacteria'.

An important shift in the foraging landscape happens about now. In 2019, neatly enough, the bees abandoned the May blossom of hawthorn at the end of May. In other years they carry on collecting its pollen into early June, but no later, and this marks the end of Synge's spring – her second division of the pollen year. In the next section, which she called midsummer, bees move from the field edges into open ground, from trees and shrubs to herbaceous plants, and the pollen that they collect can vary greatly from place to place, and sometimes depend almost entirely on the crops that are grown there.

June 2019, Week 2

Predominant		
Secondary	*Ceanothus*	California lilac
Important Minor	*Rosa*	
	Taraxacum type	
	Meconopsis cambrica	Welsh poppy
	Papaver rhoeas	Common poppy
	Heracleum spondylium	Hogweed
	Sambucus nigra	Elder
	Pyracantha	Firethorn
Minor	*Polemonium caeruleum*	Jacob's ladder
	Plantago	Plantain
	Cornus	Dogwood
	Trifolium	Clover
	Rubus	Bramble
	Rumex	Dock
	Symphytum	Comfrey

In our heavily-modified landscape, the kinds of plants that we choose or allow to grow have an enormous impact on honey bees at this time of year. A native *Cornus*, or dogwood, grows in hedges, but other varieties are planted widely as shrubs and often form part of the block planting around office developments or new housing. Its dark yellow pollen can appear at this time of year, and may be as likely to come from this kind of landscape planting as from a hedge. The grains are about 60μm across, and have unusual furrows, which stay at much the same width all the way along, making them look a little like a watch strap.

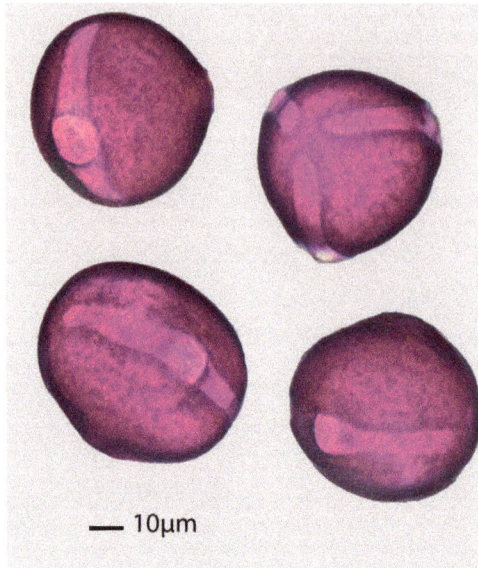

Figure 47: *Cornus.* Family: Cornaceae (dogwoods)

Another shrub from the same kinds of habitat is *Pyracantha,* or firethorn. It is a member of the rose family, and the pollen grains have the usual shape, but they are a distinctive dark green and have narrow ends to the furrows, stretching up to the pole.

Figure 48: *Pyracantha.* Family: Rosaceae

Some plants, like these two shrubs, have become more widespread because of garden or landscape planting, but so much of our non-urban landscape is given over to agriculture that it is the plants there which have the greatest effect. 'The fields studded with the useless daisy are a desert to the bee,' Robert Huish complained in 1844, 'but it is the fields which are whitened with the buck-wheat, the plains which are gilded with the flower of the wild mustard, the turnip, and the whole of the brassica tribe, that furnish the bees with a continual supply of food.' A hundred years later, Percival and Synge both recorded heavy collection from the clover of pasture land, arable weeds like charlock, and the field bean, *Vicia faba*, which was an important crop around both study sites. Today, as we have seen, oilseed rape is significant for a few weeks in lowland regions, but many of the cropfield weeds have become rare, and another of Huish's worries, that 'a highly cultivated country is by no means beneficial to the bee, for as soon as the harvests are got in, the fields are a complete desert to the bee', has become an important concern. An attempt was even made in 2016 to measure the total amount of nectar that plants provide, and how this has changed over the years for which numbers are available. Between 1978 and 2007, land use in Britain was surveyed at regular intervals in the Countryside Surveys, and used to follow changes in different habitat types. The nectar produced by samples of the most common plants in those habitats can be measured, and these were then used in the 2016 study to estimate how many tonnes of sugar the plants provide.

This only takes us back about forty years, but in the 1930s another survey had been done, and this can give us an idea of the changes which have occurred over a longer timescale. Under the direction of L. Dudley Stamp, of the London School of Economics, volunteers colour-coded maps according to broad categories of land use. In this early example of citizen science, Stamp estimated that a quarter of a million schoolchildren had been involved in the work, and he was far from idle too. 'I must have covered thousands of miles myself,' he wrote in 1948, 'often standing up on the front seat of my car with my head through the sunshine roof and a roll of six-inch maps in front of me' – an image that appears less frightening when he tells us that it was his wife who was driving. Nothing illustrates the impact of technology better than these surveys. It took fifteen years to complete and publish Stamp's work, although most of the fieldwork was done by 1934 and the rest of the process was slowed by the war, including the loss of all the printing plates in an air raid. Stamp himself noted that nothing like it had been done for 850 years, since William the Conqueror surveyed his new home, and I can remember seeing the huge, colourful sheets of Stamp's maps still being used fifty years after his survey was done. He would have recognised the methods used in 1978, as people on the ground recorded what they saw, but already then the data was being analysed by computer. Soon, satellites were photographing the survey sites, and the remote images were used alongside the human records to train the computer to recognise vegetation types. Before long, we can expect automatic interpretation of satellite pictures, and any new conqueror could have a survey in a week.

The results from the first Countryside Survey of 1978 were gloomy compared with the picture from Stamp's time. The post-war transformation of agriculture meant more food for us, but less nectar for insects. Across England and Wales (the only two countries covered by Stamp) the estimated nectar supply had decreased by 32%. The surveys after 1978, though, suggest an increase, to which the spread of oilseed rape has contributed. The problem then, as Huish noted, is what happens outside the few weeks when the fields flower. A large-scale study carried out recently in an agricultural area of France, where the landscape was dominated by fields of sunflowers and then maize, found that there were so few nectar sources other than when these crops flowered that many beekeepers needed to feed their bees with sugar syrup in the weeks between.

One answer is to share the land. Another French survey of pollen recorded bees using oilseed rape early on, then sunflowers later, but supporting themselves on semi-natural habitats like woodland between the flowering of the crops. Where EU funds were once used to encourage the spread of oilseed rape, some of that money now supports the planting of arable land with insect food as well – sources of nectar and pollen that will feed insects over a longer season than a single crop can. The 2016 nectar assessment suggested that the contribution from schemes like this was low so far, considered on a national scale, but they do offer one solution. Another is to look at the land that the Countryside Surveys record as a type called improved grassland. This covers almost a third of the UK, and nearly half of Wales, and so any change to its value for insects can have a huge effect. 'Improved' means more productive – more grass over a growing season – and this is often achieved by applying fertiliser and sowing with species that respond to that fertiliser with vigorous growth. In many case, returns are maximised by frequent cutting for silage. The result is sometimes fields dominated by two or three kinds of grass, particularly rye-grass, and little else, certainly not for bees.

If there is no flowering crop to dominate the landscape, then bees must collect wherever they can, and the number of different types collected tends to reach its maximum at this time of the year. Garden plants rarely supply enough pollen to be more than minor sources, and in some years Jacob's ladder, *Polemonium caeruleum* isn't recorded at all. But in 2019 a little was collected, and the grains are striking, with a striate surface and large numbers of pores.

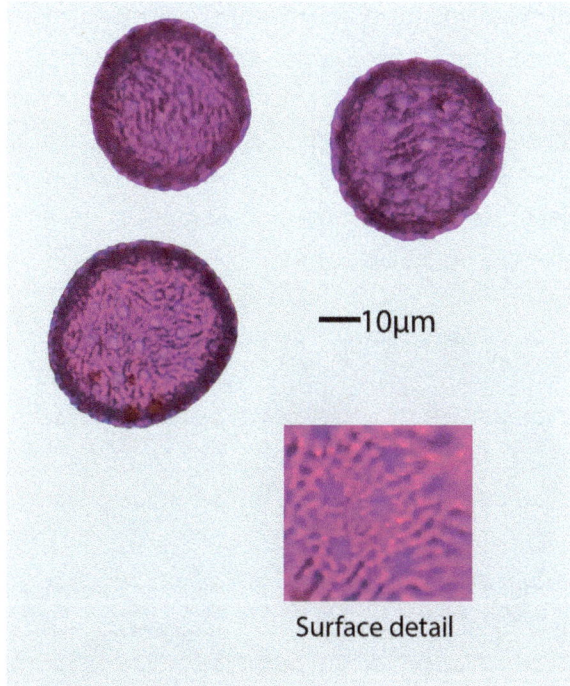

Figure 49: *Polemonium caeruleum*. Family: Polemoniaceae (phlox)

Bees also continue to take advantage of plants which are wind pollinated, and have nothing to gain from insect visits. Docks are common on rough grazing and waste ground at this time of year, and their pollen is sometimes collected in small amounts. The furrows are unusually narrow and long, and the shape is transferred to the pores, which look as though they have been stretched towards the poles.

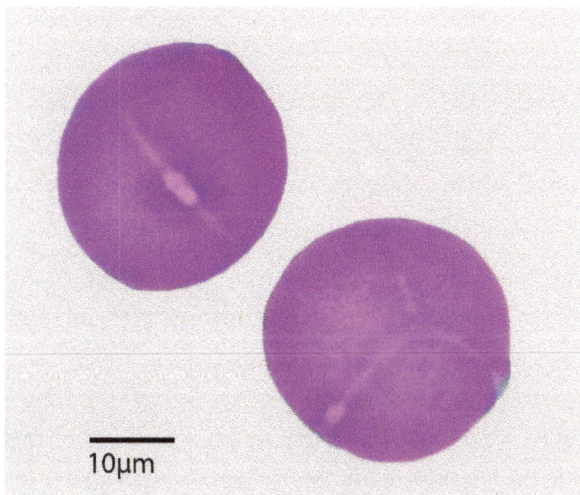

Figure 50: *Rumex*. Family: Polygonaceae (knotweeds)

The docks are growing amongst grass, and grasses are plants that dominate the open spaces of the world, from prairie to steppe. These are conditions which favour wind pollination, and grasses are angiosperms which have turned back to the wind, but honey bees gather yellow pollen loads from them in small amounts from now until the maize crop late in August. The pollen of the different types is too similar to allow most to be identified: they all tend to be round, with the smooth surface you would expect of a wind-borne grain, interrupted by only a single pore.

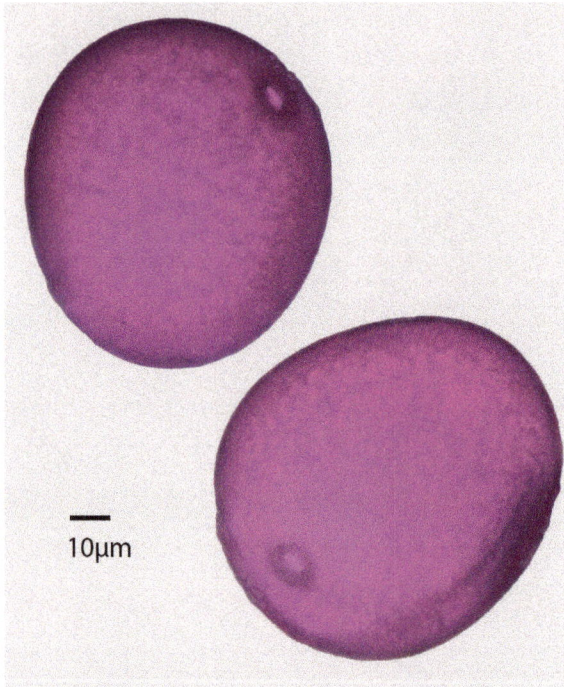

Figure 51: Graminaceae family (grasses)

Sometimes, size or flowering time can give a clue as to which grasses the bees have visited. Most of the common grasses are not flowering yet, and this pollen may well come from the meadow foxtail, *Alopecurus pratensis,* which is the most common type in flower now in the field nearest to the hives.

Much more recognisable are the pollen loads from the red poppy, *Papaver rhoeas.* This would have been a common agricultural weed at one time, and is still significant in agricultural landscapes where it does occur. It also seeds readily on waste ground, and bees bring from it pollen loads which are almost black, and different from anything else collected at this time of year. Analysis of the pollen structures on the plant shows that they absorb the ultraviolet wavelengths which are invisble to us but detected by the eyes of bees, and to a bee they must contrast dramatically with the petals (red to us, of course) and the female parts of the flower, all of which strongly reflect UV light. The poppy offers no nectar to insects, only pollen, and is advertising clearly where this reward is to be found.

Under the microscope the grains show a clear pattern of granules scattered over each of the furrows. These are broad, and have no pores.

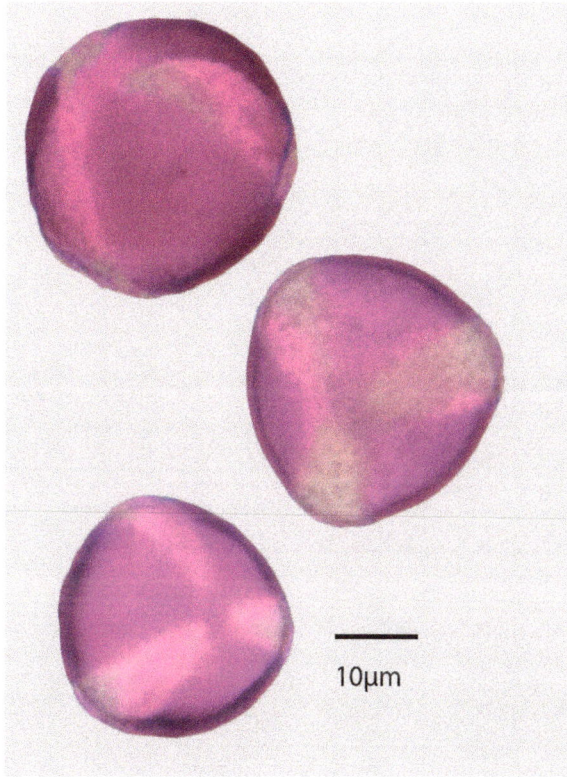

Figure 52: *Papaver rhoeas.* Family: Papaveraceae (poppies)

The general appearance of the grains is like those of the Welsh poppy with which we began this section, and both are alike in the flat, open flower surface which they offer to the bee. As a result of his observations on pollen collection, Parker placed flowers into different groups according to how their pollen was presented. The poppies are examples of his 'open flowers', where the insect can simply run across the surface, gathering pollen with mouth and forelegs, taking to the air now and then to pack the grains into their baskets. He also grouped together plants with spikes or catkins, like hazel, willow or oak, where he described bees moving up from the base, again leaving the flower to shape the pollen pellet before returning or moving on to another bloom. Later in the year we will see Parker's closed flowers, like white clover, which the bee needs to force open with her forelegs before inserting her head and using her mouth and forelegs to gather the pollen. Later still, fuchsias will be an example of his tubular flowers, where the structure only allows the bee to insert her proboscis in search of nectar. Here there is no room to move around collecting, and pollen probably only rubs on to her head rather than being deliberately gathered, and is then later groomed off by the legs.

Darwin proposed that differences like these in flower structure may at least begin to explain the constancy of honey bees to a particular species. Once they know how to collect pollen from a particular type, the argument goes, it is more efficient for them to seek out flowers of that kind rather than visit a new one and waste time puzzling over how to handle it. It has the efficiency of the worker on an assembly line, repeating a single, specialised task, and as long as there are enough flowers of a particular type to satisfy it, a honey bee will stick to those. But poppies suggest that this can't be the whole story. In lowland arable areas, before more intensive farming reduced their numbers, it was common to find several species of *Papaver* growing together, and a study in the late 1950s near Oxford watched honey bees visiting them for pollen. Although the five types growing there could hybridise with each other, they rarely did. This was partly because some of the species had mechanisms that allowed them to self-pollinate and exclude pollen from other plants, but also because the honey bees that were their main pollinators stuck mostly to a single species for their visits. Yet the different types were similar in structure, and there was no obvious reason, as far as pollen gathering went, why the bees should discriminate between them at all. Perhaps, though, even subtle differences are important, not because they affect how a bee manipulates a flower, but how it recognises it as the source of food. Some have argued that the bee's memory, and its limitations for storing what represents the food source, may then be part of the story. Poppy species that look similar to us may be more distinct to a bee, and foraging might be less efficient if she tries to remember and recognise more than one. A bee may ignore all but the current target because, in a way, she doesn't see that they are there. The truth is that although we may be able to follow the mechanics of pollen gathering, the bees we can see working so busily are still behaving in ways that we don't yet fully understand.

Poppies make their pollen easy to collect. Other flower structures present more of a problem. For each of Parker's different types of flower structure, though, and the many variations on them, there are at least some kinds that honey bees can use. With the swift, co-ordinated movements of mouthparts and legs that Sladen and Casteel deciphered a hundred years ago, they can extract and pack pollen from them all. They must have pollen, and its importance is shown by the behaviours and the external structures that they use to collect it. Some of these are found also in bees that live alone or, like bumblebees, in small groups. But one aspect of foraging is unique to the societies of honey bees, and that is what we will look at next.

6. Searching and dancing. June 2019.

Of all the wonderful structures and behaviours that are found amongst insects, nothing has inspired more study than the complex societies that a few types, honey bees among them, have evolved. They remind us of our own networks, and many have looked for analogies between insect societies and our own and offered the termite mound or the hive as a model of harmony and order. The title of the first English-language book on beekeeping, '*The Feminine Monarchie*', is a description of social arrangement. It was one of the first accounts to confirm that the hive revolved around a queen rather than a male insect, and perhaps it was no accident that the author, Charles Butler, lived the first half of his life during the reign of Elizabeth. We are a long way from fully understanding the evolution and organisation of insect societies, but as far as pollen is concerned, we can look at them in two ways. Firstly, they are a challenge – how can so many mouths be fed ? Particularly when, as we have seen in the year so far, the target moves as the seasons change. Sometimes there are trees which offer rich rewards, but may be scattered in the landscape. Sometimes there are hedges, and sometimes crops which appear here and there. Sometimes there are no more than patches of flowers or garden shrubs. All of these have to be found, and enough of them to raise thousands of bees in every hive. That may mean foraging over an enormous area. But the second way to consider the matter is that the society which creates the problem also offers the answer. If there are lots of bees to feed, there are also plenty to search for food. And in their searching, they show clearly how they form a social group rather than a collection of individuals huddled in the same hive. Rather than each foraging independently of the others, they share information about what they have found, and once one bee locates a good source of food, others can be recruited to exploit it. When every flower on a tree seems to hum, it's not because hundreds of bees have each found it on their own: most have probably been told where it is.

Aristotle is again our first source for this idea of recruitment. He wrote that food near a colony might be undiscovered for a while, but that once one bee had found it, many would soon arrive. Until the 20th century, how this might happen remained a mystery. Or even whether it did: as late as 1882, Sir John Lubbock, in his *Ants, Bees and Wasps*, doubted whether bees recruited one another to a food source. A very Victorian polymath, Lubbock was a distinguished banker who made notable contributions to archaeology as well as biology, the latter stimulated by a friendship with Darwin. He recorded on several occasions the same bee returning again and again to a source of honey – 'a perfect Eldorado' – as he described it, but with no sign of recruitment. 'Some bees, at any rate,' he concluded, 'do not communicate with their sisters.' But to Maurice Maeterlinck, a Belgian who matched Lubbock in his range of interests, being a poet and playwright as well as a beekeeper, it seemed obvious that social life must include communication. 'There can be no doubting that they understand each other,' he wrote in *The Life of the Bee*, in 1911, 'and indeed it were surely impossible for a republic so considerable, wherein the labours are so varied and so marvellously combined, to subsist amid the silence and spiritual isolation of so many thousand creatures.' He believed that they could share all kinds of information:

> For the mutual understanding of the bees is not confined to their habitual labours; the extraordinary also has a name and place in their language; as is proved by the manner in which news, good or bad, normal or supernatural, will at once spread in the hive; the loss or return of the mother, for instance, the entrance of an enemy, the intrusion of a strange queen, the approach of a band of marauders, the discovery of treasure, etc. And so characteristic is their attitude, so essentially different their murmur at each of these special events, that the experienced apiarist can without difficulty tell what is troubling the crowd that moves distractedly to and fro in the shadow.

And although Maeterlinck admitted that not every bee which found a food source would bring others to it, on average, he thought, four bees out of ten would do so. Some were particularly persuasive:

> I even one day came across an extraordinary little Italian bee, whose belt I had marked with a touch of blue paint. In her second trip she brought two of her sisters, whom I imprisoned, without interfering with her. She departed once more, and this time returned with three friends, whom I again confined, and so till the end of the afternoon, when, counting my prisoners, I found that she had told the news to no less than eighteen bees.

As Maeterlinck suspected, all kinds of information passes between members of the colony along a network of sensory cues, some of which are still little understood. As far as pollen and nectar are concerned, the most widely-known are the bee dances, decoded by Karl von Frisch between 1920 and the late 1940s. The curious repetitive movements that some foraging bees make following their return to the hive had by then been noticed many times. The 19th century inventor of the modern hive with its removable frames, L L Langstroth, was one example. He described how a bee, having returned to the hive with a pollen load, might shake its body in a characteristic way. Nicholas Unhoch reported the same kinds of movements in 1823, but without understanding their role. Both probably thought that the movement did no more than convey a general excitement that food had been found. Something like this occurs in bumblebees, which don't show direction in their 'dances', but rather just run up and down. The signal is simply that there is something exciting out there, and their nestmates are thought to find it for themselves, using the scents that the agitated bee carries back from its foraging site. But some observers suspected that honey bees were different. Thirty years before Unhoch, a German pastor, Spitzner, had described bees returning from a food source to his glass-walled observation hive, and there twisting in circles in front of their nestmates. 'Full of joy', he concluded, they were conveying the scent of the food, and would then lead a crowd of followers back to what they had found. This must have seemed the likeliest mechanism for recruitment: that the pioneer would directly lead others back to the source. With honey bees, however, von Frisch was able to interpret what he termed the *Tanzsprache* or 'dance language' of the bees to show that the movements contained messages that, rather than recruiting other bees to follow the dancer, told them where to find it for themselves. His conclusion was that the dance conveyed information about the distance and direction of the forage that the dancing bee had found, and also an indication of its value. Other bees might then be stimulated to look for the source, and before long a flowering tree found by one bee could be thronged with workers from the same hive. This co-operative foraging is such an efficient way of combing a landscape for pollen, nectar or water that it has been suggested as one of the main factors driving the whole evolution of insect social life.

von Frisch published an English-language review of his and his students' work on bee dances in 1967, and his discoveries later won him a Nobel prize. Before that, it had saved his career and perhaps even his life. The Nazi regime discovered that he had a Jewish grandmother, but von Frisch was able to continue working, partly because there was concern in the 1940s that *Nosema* infection threatened bee populations and that research like his was important if wartime Germany was to grow enough food.

In his early publications in the 1920s, von Frisch described two distinct types of dance, each attracting an audience of followers. The first was the round dance where a bee ran in circles, first in one direction, then the other, and he believed that this was performed by nectar foragers. The second has become known as the waggle dance, where 'waggle' refers to the wagging movements, particularly of the abdomen, that the bee makes during a short run along the honeycomb. These were the agitated movements described by Spitzner and Unhoch, and von Frisch thought at first that they belonged to pollen foragers, allowing them to beat the faces and antennae of dance followers with the pollen loads that the forager carried. The movements are accompanied by a buzzing sound, and at the end of the run the bee moves in a semicircle back to the starting point.

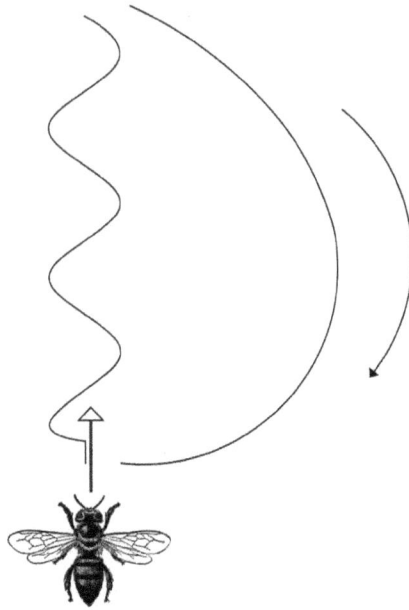

Figure 53: waggle dance first stage

She then repeats the waggle run, followed by a semicircle on the other side.

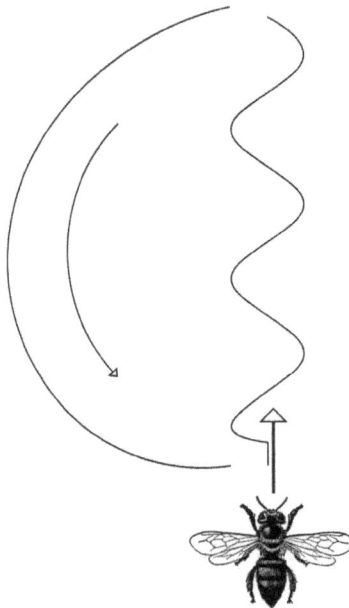

Figure 54: waggle dance second stage

These circuits are repeated, and the dancing bee may draw the attention of more and more nestmates, potential recruits to the foraging site that she is reporting. Now and then she stops and feeds these followers from her honey stomach, and then begins to dance again.

It soon became clear that there weren't different dances for different kinds of food. O W Park noted in the *American Bee Journal* for 1923 that he had seen waggle dances by bees collecting water, and by nectar foragers too. Over the next twenty years, von Frisch continued to study them, and gradually concluded that the waggle dance was something unsuspected for an insect: a sophisticated means of conveying information about what a forager had found so that the whole colony could use it.

If we think first about the distance of the food source, the dance followers learn this from a number of elements. The waggle run is longer for greater distances, and the duration of shaking and buzzing increases. We have learned to interpret the dance from what we see, but for the bees it happens in the dark, and they may follow it mostly with their antennae, sensing the vibrations caused by the movements and the sounds. The overall tempo of the dance is also important, with a slower dance - fewer circuits completed in a given time - for resources that are further away.

These distance elements also seem to convey a sense of the quality of the pollen or nectar source. Dances vary in the level of 'enthusiasm' of the dancer: the extravagance of the shaking, the intensity of buzzing and the number of circuits made, can all be manipulated by providing sugar solutions of different quality. The more excited the dancer, the more recruits she attracts. These then seem to respond with more excitement to the samples that she gives them, and their response in turn encourages her dance. Eventually, more recruits will set out for the source that she has found than for a source advertised by a lacklustre dance. The level of response also depends on the food stores of the colony. In desperate times, workers will respond to news of even a poor-quality source: a site where there are few flowers, perhaps, or far away. But when the colony has plenty of food, only news of something really special will get the workers out of bed.

The recruit now has an idea of how far to go, and has perhaps also taken in the scent of the source, to give an idea of what she is looking for. All that remains is direction, and the waggle dance conveys that too. There are some kinds of bee that dance on horizontal surfaces, like a branch above a hanging nest. On this kind of stage, the direction of the food source can simply be the direction of the waggle run. The line of the run points to it, by setting out the angle to the sun that the recruit should follow. On a vertical comb, though, inside a tree or a hive, our honey bees need another way, and here gravity is used as a reference point. A waggle run going vertically up, as in the diagrams above, means that the food is directly towards the sun. A run at a particular angle to the vertical tells the recruits to go in a direction at that angle from the sun. Even on overcast days when the sun is hidden, it still creates a pattern of polarised light, and although our eyes can't detect this, the bees can, and use it to interpret the information in the dance.

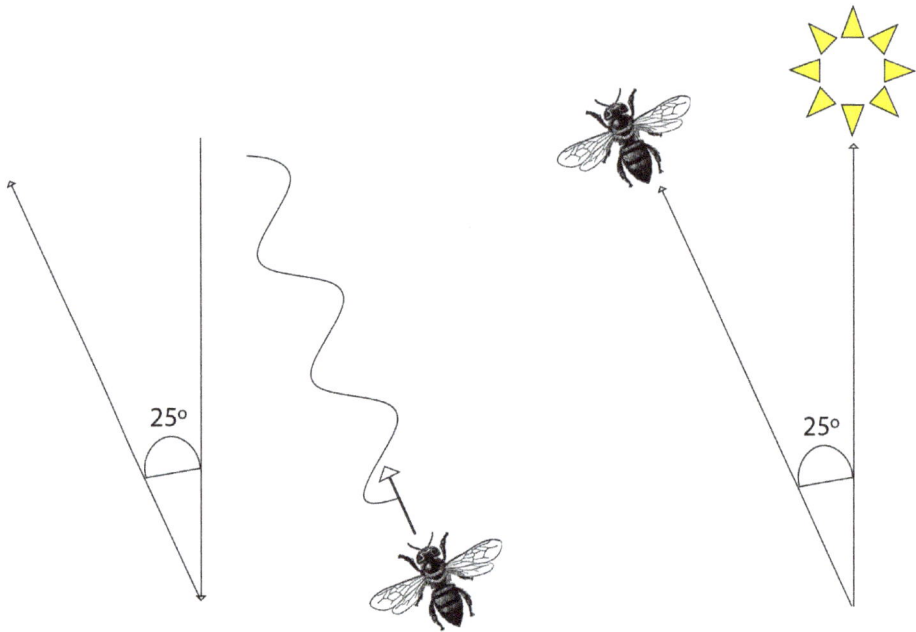

Figure 55: Inside the hive, a bee makes a waggle run at 25° to vertical (left).
A recruit leaves the hive and flies at 25° to the sun (right)

Although it was clear that round dance and waggle dance were not exclusive to particular kinds of forager, von Frisch continued to believe that they were different kinds of performance. In contrast with the waggle dance and its multiple components, the round dance would simply encourage other bees to leave the hive, and then look for the source, rather like bumblebees. He proposed that the honey bee round dance signalled food close enough for scent information to be enough to find the source, and that the waggle dance was something different: a more sophisticated kind of information transfer. More recent experiments, however, have suggested that the round dance does contain information about where the dancer has come from, and so may be best thought of as a truncated version of the waggle dance, rather than a separate type.

These are relatively minor modifications to what von Frisch had found. More seriously as a challenge, some went back to the idea that the dances communicate no more than the general excitement that bumblebees transmit, and argued instead that the real message lay in the scents which the dancer carried back to the hive. There is no doubt that communication involves more than just the dance movements: von Frisch himself had shown that the followers responded to scent and to the taste of the samples fed to them by the dancer. The question became what information lay in the dance itself. The controversy went on for some years, but eventually it became clear that von Frisch was right, and that the dancers do communicate information through their dance about where to find the source of food. The intensity and direction of the movements are part of the communication. It is only one component, though, of the information that Maeterlinck imagined flowing through the colony, and the balance between dance, scent and taste is still being worked out. Using techniques developed to look at how information moves around social networks, for example, a 2020 study in London concluded that workers recruited to new sources of food were led to them by dances. Foragers stimulated to return to known sites, on the other hand, perhaps after a few days of bad weather had kept them in the hive, were responding to scent and to samples of food. Few would argue now with von Frisch's claim that the dance language is the most remarkable means of communication found in anything other than primates, but it is not the only method that honey bees use.

Today, the dances are so well-studied that they have become research tools in themselves: investigators record and analyse them to discover where bees are feeding, and how far they go. And yet we are still some way from understanding exactly how the followers which gather around a dancing bee decode the message in the dance. Some studies have even shown that dance information makes no difference to how well the colony collects food. These experiments involve hives turned on their sides so that the honeycomb lies horizontally rather than standing vertically as it would normally do. Without a vertical comb to indicate gravity, the dancing bee can no longer show the direction of the food source. The hives can then be weighed to measure the effect of this disruption on the collection of food. One factor which seems to affect the results is the kind of landscape in which the bees live, and this points out a weakness in the early studies of dances. In these, the usual method was to set out a source of sugar syrup at different distances and directions from the hive. This made sense when the intention was to relate the dance to a known source of food, but of course the environment in which the bees normally forage is much more complex than that. Some have even wondered whether the dances evolved amongst the vegetation of tropical landscapes, and may not be as valuable in modern agricultural regions of Europe and North America, with their less-complex arrangements of plants.

The ecological context has certainly played a role in how the dance has evolved. Some studies have investigated whether sound might be a component of the message about food. Even if it turns out that the sound of the dancer does no more than attract attention to the dance, it is easy to see how sound might be particularly significant for the kinds of bee that live in the darkness of a hive – or a cave, or tree. In contrast, some types of Asian bee that make their nest in the open seem to instead use particular postures as part of their dance, something which would be quite useless in the dark.

A further invisible component of the dance is the scent that dancing bees release. As we have seen, foragers carry odours from the environment, including some from the flowers from which they have been feeding, and these, it is thought, may help recruited bees to zero in on the source once the dance has taken them in the right direction. But they also, when dancing, release scents not picked up outside, but produced within their bodies. These chemicals alone, synthesised and released into the hive, increase the number of bees which fly out. Presumably they are stimulated to look for food, even if they haven't seen a dance to tell them where to go.

The fact that bees could use a dance to tell their nestmates about where to find food is so astonishing that there may have been a temptation in the early days to imagine that the dance itself was all there was to it. Like robots programmed with their instructions, foragers responded blindly to the information they received. This is not the case, and individual bees are far more flexible than that. von Frisch himself, noting that bees were more likely to follow dances to places they already knew, seems to have understood that this was so. Bees have even been shown to ignore dances that tell them about food sources much closer than the ones they are already using, as long as these sources of their own continue to satisfy them. Two thousand years on from Aristotle, we are beginning to see that recruitment of foragers is affected by the landscape in which the bees live, and involves a mixture of the social knowledge that dancing lets them share, and private knowledge of their own. Bees respond to dances, but in a way that depends on their own knowledge about the plants around the hive, and the memories that they have made during their short foraging lives. And the dancers are flexible too: foragers will make less effort to recruit other bees if a dead bee is placed near the flowers they have found, perhaps so that they don't lead their nestmates into danger. The motivations and impulses that feed into the decisions of these insects begin to seem little less complex than our own.

Around my apiary, some decide to visit the hogweed flowering in June on field margins and road verges. The yellow pollen has long grains with three narrow furrows, and is typical of the umbellifer family to which they belong. The flower structure is typical too, with many tiny flowers creating a flat disc, all supported by short stalks spreading from a single point, like the ribs of an umbrella. It is this resemblance which leads to the name given to this type of flower structure – an umbel – and the disc is a landing platform for many kinds of insects as well as honey bees.

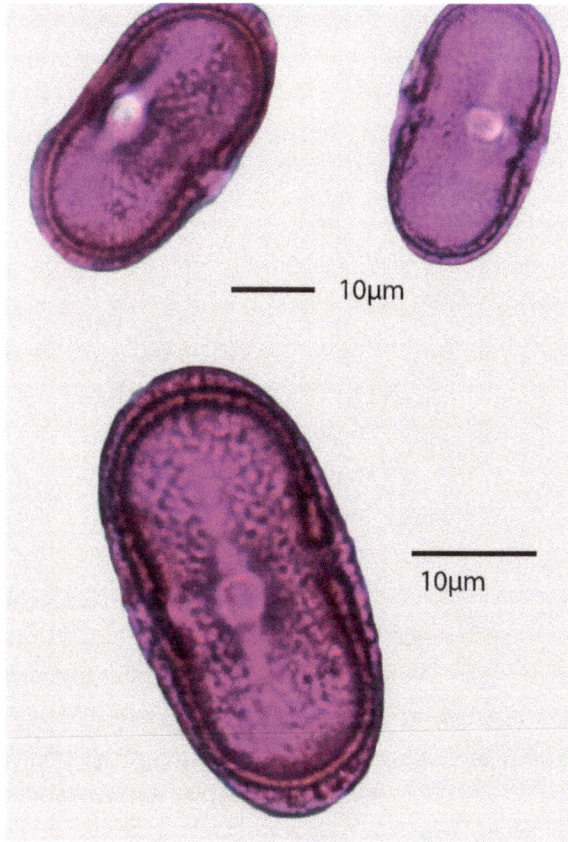

Figure 56: *Heracleum spondylium.* Family: Apiaceae (carrot, parsley)

A family resemblance is clear in another member, cow parsley *(Anthriscus sylvestris)* which has been flowering for some time on every roadside. The grains are smaller, but otherwise much the same. For some reason, though, the bees here collect pollen from hogweed for several weeks in quite significant amounts, but from cow parsley hardly at all. We may know a lot about how bees convey information, but there is still much to learn about how they decide to use it.

One of the most surprising discoveries that emerged as researchers began to follow up on von Frisch's work was that the dances were imprecise. As the dancing bee repeats her steps, the waggle angle varies, which means that the followers have a range of directions that they can take, rather than a precise angle relative to the sun. A further element was also reported in the 1950s: that the dances become more precise – that is, the repeats are more similar to each other – when the food source is further away. Ever since, researchers have wondered about two possible interpretations of these findings. Are dances imprecise because the bees can't do any better, or is there is some advantage in it, meaning that the instructions are fuzzy for a reason ?

The original conclusion was that the bees are as precise as they need to be. Imagine you set off from your house at an angle of 26°, and your friend sets their compass at 25°. Over a short distance, this has little effect, and the two of you will remain close together. If you were aiming for a tree, both of you would probably get close enough. If you carry on moving away from the house though, eventually a small difference in the starting angle will take you so far apart that one of you will miss the destination. So the idea was that when the food source was a long way away, and it was important to get the angle right, the bees made their dances more precise.

This suggests a certain sloppiness in dance behaviour: the bees are like performers who raise their game for the big stage, and get it right only when they have to. A later proposal was more flattering: that the imprecision was there on purpose, and its function was to spread the foragers out as they left the hive. After all, the intention is probably not to direct followers to a single, precisely-located flower, but to a general place where trees, a hedge, or a field are offering rewards. This could also explain the effect of distance – that the dances become more precise when the pollen or the nectar is further away. Foragers following a dance to a nearby location will spread out widely, because they will follow a range of dance angles. Meanwhile those responding to a dance for somewhere far away will start off in very similar directions. But not exactly the same, so that over the greater distance they will eventually fan out too, and the overall effect might be to end up with a similar spread of foragers, no matter how far they go. In this elegant interpretation, the recruited bees are like a net cast out over the landscape, and the function of the imprecise dance is to keep the net the same size, however far it is thrown.

Those who favour this proposal compare food dances with the other occasion in which bees dance to show direction. This is when the colony divides through swarming, with some bees left to raise a new queen in the original home and some guiding the old queen to set up in a new site. Before the journey begins, some bees have been searching for suitable places, and once the swarm has left its hive it settles nearby, and these scouts communicate what they have found through a dance. It may even be that this was the original function of dance behaviour, and its use to communicate food evolved later. Somehow, the swarm evaluates the various proposals, makes its decision, and flies away. Now all the bees have to travel to a single point, and there is no advantage in imprecision. And, indeed, these nest dances show much less variation than the dances for food: proof, this line of thinking goes, that dancing can be precise, and if it isn't, then it isn't for a reason.

The problem with comparing nest dances and food dances is that they take place under different conditions. Inside the hive, a bee stands on a wax comb and dances in complete darkness. The sun can't be seen, so is referred to indirectly. Out in the swarm, the dancer stands on other bees and in full view of the sun. Perhaps then, a third view of dances goes, they are imprecise not for a reason, or because they don't have to be better, but because under the difficult conditions of the hive, this is the best the bees can do. To support this idea, we can note that bees usually follow more than one dance before they set out, perhaps attempting to work out a middle way between the various angles proposed. Or we can consider reports that the more an individual bee changes her dance between each repeat, the more of these dances a follower will watch before she leaves, as if trying to make sense of a message that isn't clear. And what about the point that dances for more distant locations are more precise ? This may be because long-distance dances involve a longer vertical run. A longer run gives the bee more time to adjust its gravity reference, and so give a more precise sun angle to be followed. At the moment, we can only say that when bees pass on information about where food is to be found, they repeat their message and vary it slightly each time. Whether this has any meaning, and what that might be, isn't clear. More than seventy years after von Frisch began to decode the bee dance, this is another thing still to be learned.

7. The pollen landscape. June/July 2019.

We are now well into the 'June gap'. The harvest from trees, shrubs and oilseed rape is over, and the main season for summer flowers like bramble and clover only just begun. Hives sometimes lose weight at this time of year, as their increased populations come up against a shortage of food, and they turn to their stores to tide them over. With no single widespread pollen source, the samples contain many different kinds, and change from week to week. Privet, fashionable once and still common as a hedge around older houses in this area, comes into flower. Its scent is clear on a still day, and no doubt obvious to bees, which visit the hedges for both nectar and the yellow pollen. The nectar is said to ripen to a honey with an unpleasant taste, and the pollen grains are distinctive too, with a clear net across the surface, and a sail-like fringe around the edge. The furrows can be indistinct, but the shape of the grain makes clear that there are three of them, each without a pore.

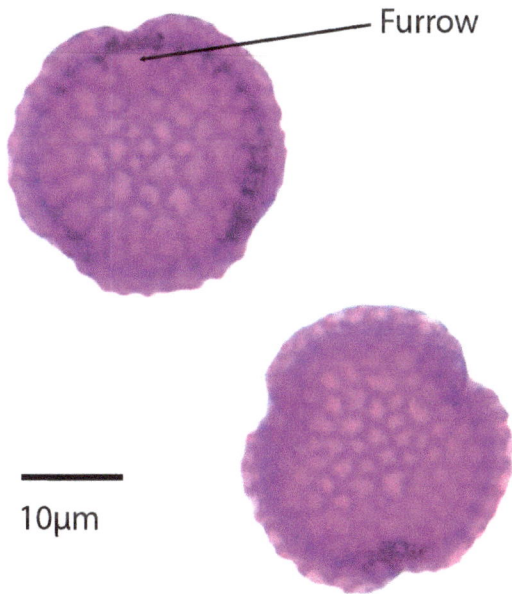

Figure 57: *Ligustrum vulgare.* Family: Oleaceae (olives)

A shortage of pollen sources might also explain the appearance of some bright yellow pellets in the samples, which look at first like normal pollen loads, but under the microscope are very different.

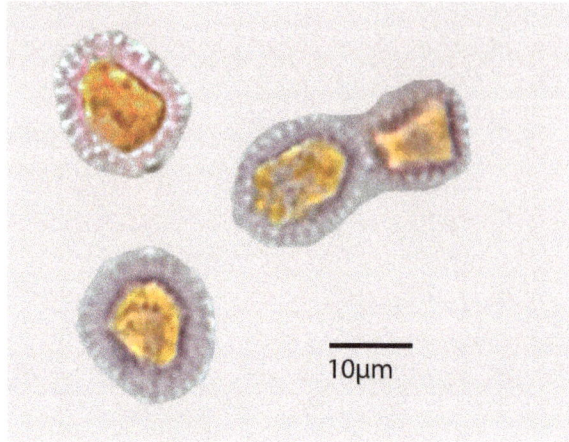

Figure 58: Fungal spores.

They don't absorb the stain which turns pollen grains pink, and with their yellow centre and translucent fringe are sometimes described as being like fried eggs. They are not from plants at all, but fungal spores, which honey bees collect from a number of different types of fungi, gathering them into pellets exactly as they do with pollen grains. The behaviour was first recorded by the Reverend J L Zabriskie, a beekeeper, but a keen naturalist and microscopist too. At one time president of the New York Entomological Society, Reverend Zabriskie discovered two species of fly, parasites on a solitary bee, which now bear his name. In 1875, he wrote in the *Beekeepers Magazine* that he had seen bees returning to the hive carrying loads of a bright vermilion, which under the microscope turned out to be the spores of a rust fungus that infects raspberries.

June 2019, Week 3

Predominant		
Secondary	Trifolium	Clover
	Brassica	
	Rubus	Bramble
Important Minor	Rosa	
	Meconopsis cambrica	Welsh poppy
Minor	Papaver rhoeas	Common poppy
	Heracleum spondylium	Hogweed
	Ligustrum	Privet
	Graminae	Grass
		Fungal spores
	Ceanothus	California lilac

Then, in February 1901, W H Lang of the University of Texas observed bees actually collecting fungal spores, this time from orange rust fungus on blackberry plants. Lang seems to have been unaware of the earlier work, writing that for him this was a 'hitherto unknown habit', but in any case raised for the first time two important points about what the bees were doing. Firstly, he noted that there were flowering plants in bloom at the time, and that the bees seemed to be gathering fungal spores by preference. Secondly, he wondered whether honey bees might be one way in which fungal disease is spread. We are still not sure about whether spores have some particular nutritional value, or whether bees collect them only when there isn't enough pollen to be had. A pollen shortage was Mary Percival's view, when she found her bees bringing in almost 3000 loads of rust spores over a few days in August. And bees will collect all sorts of powdery materials, from sawdust to coal dust, if there is no pollen to be had. On the other hand, analysis of fungal spores has suggested that they can have protein content comparable to that of some types of pollen, and a variety of fatty acids too. Perhaps Lang, who concluded that 'the honey bees know their business, and would certainly not gather the spores if they could not use them', was right. As for transmitting disease, it seems certain that honey bees are one of the many ways in which fungal spores are spread. An example was reported recently in New Zealand, where high prices have encouraged the production of manuka honey. The movement of bee hives to remote areas in order to make the honey was noted as a risk factor in the spread of myrtle rust fungus, the spores of which were collected by the bees.

Amongst these minor sources, the latest table also shows two plants which will become a reliable source of pollen, and of nectar too, over much of the rest of the season. The clover pollen is almost all *Trifolium repens* now - white clover - but the red-flowered species is also used, and becomes more common later in the year. The name of the second type, *Rubus,* covers a hopeless mix of sub-species and varieties, beyond all but the most dedicated specialist, and we will just think of them as brambles. In the field outside my window, there are dense clumps of these. They persist around the edge, where they are left when the hay is cut. Every few years, in a fit of over-tidiness, a tractor is sent to grub them out. The bees go without flowers for a season, but the roots persist, and soon the brambles are back. They grow in woodland too, and can flower well when treefalls let the light in. In general, though, they are persecuted, and when they thrive, do so in spite of us. Clover is treated more kindly, having been cultivated on European pasture land for at least 400 years. There the plant is so useful that breeders trying to improve the quality of grassland have devoted almost as much effort to new varieties of clover as to the grasses themselves. The purpose was to Improve the food of cattle and sheep, but bees are the most adaptable of foragers, and profited too. A survey of British beekeepers in 1952 took more than 850 samples of honey and used the pollen grains found in them to find where the bees had been feeding. The type found in the largest number of samples was clover, the nectar of which was thought to account for about three-quarters of honey produced – 'honey *par excellence*,' as F N Howes of Kew Gardens described it in 1945, 'the honey with which all other honeys are compared.' It was only in the 1980s, with the widespread planting of oilseed rape, that clover as the main source of honey was displaced.

In the field over which I am looking now, white and red clover flowers year after year. The overall mixture of plants suggests that they were once sown there as part of a seed mix for pasture, and there used to be cows there, twenty years ago. Then a new owner began to keep horses, and to manage the land for hay. Slowly, the vegetation has become more diverse, and the clover slightly less common. Meanwhile the brambles continue their guerilla existence, constantly beaten back, never quite going away. In the broader landscape too, clover has retreated. When Howes wrote in his *Plants and Beekeeping* in praise of clover honey, almost every field of grass was rich in clover, and other insect-pollinated plants too. In the decades which followed, though, the grass and clover mixes were often replaced by grass alone. Fertilisers were used to 'improve' the land: grass was cut for silage before the broadleaved plants could flower.

Our opposing views on brambles and clover, and the rise and fall of clover in grassland, are more examples of the effect that we can have on the landscape where the bees feed. Even the June gap, so well-known as to be part of beekeeping folklore, can become exaggerated, as we saw in central France in a landscape so dominated by the rape crop in May, and the sunflower crop in July that beekeepers had regularly to feed their bees in the gap between them. Equally, we can change things in a more positive way, and bee dances have turned out to be an important part of this. The dances signal what foragers have found, and as we have learned to interpret them, we have begun to discover more about how the bees understand the landscapes which we shape around them. This time of year, as the bees struggle to find enough food, is a good time to think about these places where they search, and the effect that we can have. A local example of this connection is clear in the last June sample, where brassica pollen appears again, a month after the main crop finished. A single field just across the hill, on a large dairy farm otherwise managed for silage and of little use for pollinators, has come into flower, and the bees have found it. For the next couple of weeks, it becomes their main source of pollen.

June 2019, Week 4

Predominant	*Brassica*	
Secondary	*Rubus*	
Important Minor	*Rosa*	
	Trifolium	Clover
	Aegopodium podagraria	Ground elder
Minor	*Papaver rhoeas*	Common poppy
	Rumex	Dock
	Hypericum	St John's wort
	Sambucus nigra	Elder

Meanwhile, another plant of waste ground and gardens makes a small contribution to the hive. There are several species of *Hypericum* growing either on roadsides or in gardens nearby, and they all have open flowers, with pollen easily gathered by bees. The pollen loads are a dark orange; the grains have three apertures, and they are round and unusually small.

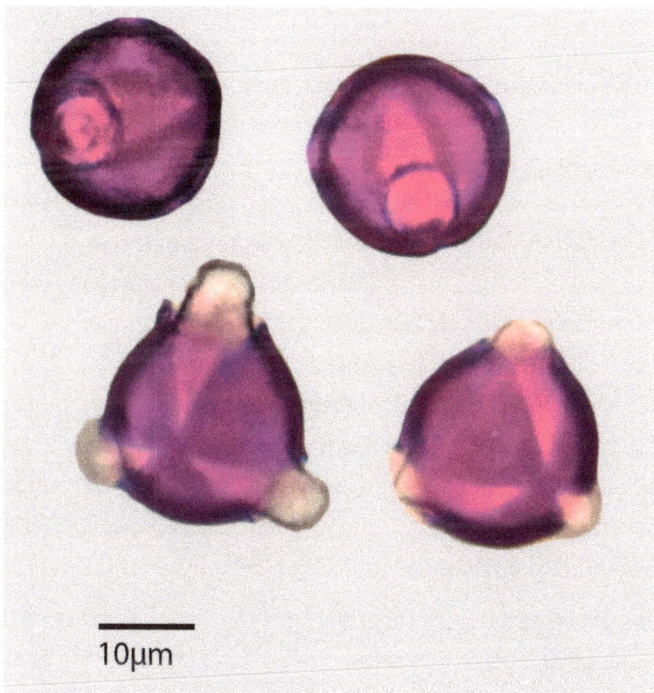

10μm

Figure 59: *Hypericum*. Family: Hypericaceae (St Joh's-worts)

There have been a number of attempts to understand how whole colonies forage, and how this is affected by the kind of habitat in which they are kept. In the 1980s, Kirk Visscher and Thomas Seeley tried to see what happened in as natural a landscape as possible, and set out hives in forest in New York state, in an area where there were no farms. What they found supported the idea of a society using dances to concentrate their efforts on the most profitable sources. Each day the bees went to a relatively small number of the patches of flowers that were in reach, and the efforts of the colony tended to shift from day to day. It is easy to imagine bees doing something similar in the tropical forests in which they seem to have evolved, where the vegetation is diverse, and patches of flowers widely scattered. But this is not the whole picture. Hives in a different habitat, this time set out among suburban gardens in Florida and California, suggested something else. Compared with the forest, flower density was much greater, and the flowers were concentrated in smaller patches. The study concluded that under these conditions, recruitment with dances was less important, and bees rely much more on individual searches. There may be two reasons for this. Firstly, the kinds of errors in the dance that we saw earlier may make the instructions too vague for followers to find the small garden patches that bees are using here. Secondly, the advantage of the dance is that it means followers can fly directly to the food and not waste energy looking for it - what ecologists call search costs. But in a suburban area, with high flower densities all around the hive, search costs are low anyway, and that benefit of following a dance is lost.

It begins to look as if colonies are flexible, adapting the way they forage to the vegetation around them. That becomes clearer with direct comparisions, where some hives are placed in one location, and others kept over the same period of time but in a different place. Something like this was tried in Germany in 2001. An agricultural landscape was chosen to represent something simple in structure – large patches of ground filled mostly with plants of the same type. This was compared with a complex landscape: a mixture of small patches of different habitat types. The bees danced much more in the second type. The landscape here may have resembled the New York forest, with patches that offered good rewards, but were difficult to find. Rather than each forager searching for herself, it may have become more efficient to follow the dance of a bee which had already found something. In the simple landscape, on the other hand, there were either empty fields, or fields full of flowers that you couldn't miss. Either way, in a simple landscape, dances would have less value, and the colony used them less.

Now, if dances are more valuable in certain kinds of landscapes, then preventing the bees from communicating with dance should have more of an effect in places like that. Another set of German experiments, in 2017, tried to look at this. Hives were once again placed in different landscapes, some more varied than others. In each of these places, some colonies were allowed to dance normally, and others turned sideways, as we saw before, to disrupt the message. Against expectations, it didn't matter what the landscape was like: there was no difference in how spoiling the dances affected foraging. As usual with bees, things aren't completely clear. In fact, one result from the study introduced a new level of complexity: when the dances were disrupted, there was much more of an effect on the collection of pollen than of nectar. The authors suggested that the bees might take nectar from anywhere, but prefer pollen from plants growing in patches of more natural vegetation, like pieces of woodland or uncultivated ground, the kinds of isolated areas best exploited through dance. We can come back to this later, when we look at the value of pollen of different kinds.

Two things do seem clear. Bees can cope with a remarkable variety of landscapes, and yet we can push them too far, and create habitats in which even they struggle to survive. To use a well-known example, for many years until the 1970s, farmers could be paid to remove hedges and create larger arable fields. The result was higher crop yields, but the loss of a habitat that offered, amongst other things, some of the pollen resources that we have seen in the year so far. The French bees, struggling to find food between their two oilseed crops, show the effect that this can have. Other creatures, less adaptable than bees, may be lost altogether. In most places, we have to shape the landscape to have places to live and food to eat, and the challenge is to do so in ways that allow other species to live there too. That means understanding how animals use those places, and bee dances can tell us something about that.

The University of Sussex at Brighton has a unit devoted to the study of social insects, particularly bees, and they have used dances almost like questionnaires, to find out what the bees think about the various forage sites on offer. Some of the farmland around the unit has been modified under one or another of the EU schemes which offer payment in return for managing land to encourage wildlife – the agri-environment schemes or AESs. Other patches of land are not farmed at all, but kept as wildlife reserves. By studying their dances, the researchers were able to see where the bees were encouraging one another to go, and therefore have an idea which kinds of land management seemed to be working best. The alternative would be to survey the wildlife in different places, but a bee colony may be able to cover an area up to 100 square kilometres, something quite beyond any normal human effort. Hopefully, areas suitable for bees would help other insects too, and so the bees could in a sense speak for them, and tell us what they would like. In Sussex, the bees showed a preference for the land managed as nature reserves, but also preferred one type of AES over another, which suggested that they could have a part to play in helping direct these kinds of landscape management schemes.

July 2019, Week 1

Predominant	*Brassica*	
Secondary	*Trifolium*	Clover
Important Minor	*Rubus*	
	Heracleum spondylium	Hogweed
	Castanea sativa	Sweet chestnut
	Aegopodium podagraria	Ground elder
Minor	*Filipendula ulmaria*	Meadowsweet
	Rosa	
	Calystegia sepium	Hedge bindweed
	Cirsium	Thistle
	Hypericum	St John's wort
	Chamaenerion angustifolium	Rosebay willowherb
	Ligustrum	Privet
	Graminae	Grass
	Rumex	Dock
	Melilotus	Melilot

One of the ways in which beekeepers adjusted to the spread of oilseed rape in Britain was to extract an early harvest of honey from the nectar of the new crop. In areas where it isn't widely grown, though, an older rhythm still holds, and the honey crop continues to take shape as the year moves on from the June gap. Pollen samples from now on show the long harvest from clover that will go on into September, and create honey which may not differ much from that made four hundred years ago, when the flower became common in our fields. This is the peak season for wild flowers, and in gardens too, and in most places, moving out of the June gap, there is no need to worry that the bees will have enough to eat. Most hedges have already been decorated with the flowers of the dog rose, and dark orange pellets of its pollen are brought in steadily, if never in huge amounts. Under the microscope, the grains have the usual shape of the family, not very different from the fruit tree pollen earlier in the year, except with a prominent central stripe of granules along each furrow.

Figure 60: *Rosa.* Family: Rosaceae

The brambles are also in the rose family, and microscopically much the same. Fortunately, the pollen has a different colour, sometimes slate-grey, sometimes with more of a greenish tinge, but always different from the orange of the rose.

Figure 61: *Rubus fruticosus.* Family: Rosaceae

Herbaceous plants, rather than trees or shrubs, will dominate the pollen landscape from now on, but two trees make an appearance late in the year. One is missing from the pollen samples here. Lime is a tree which appears late in the pollen record of recolonisation after the last ice age. This suggests that it prefers slightly warmer conditions than it would find in the north and west of Britain, and it was probably only ever in lowland England that it became a significant forest tree. But later it found favour as an ornamental, and was also planted widely in parks and along urban streets, and Howes placed it behind only clover as a nectar source for honey. It is less useful for pollen though, and Synge found it only as a minor source. The grains are unusual:

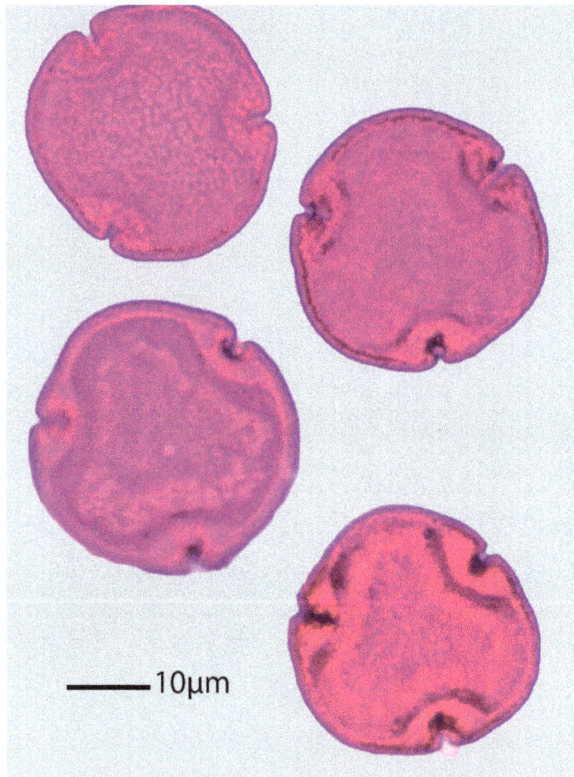

Figure 62: *Tilia.* Family: Malvaceae (mallows)

Their distinct appearance made them easy for palynologists to pick out in their samples gathered deep in waterlogged ground. They could see how lime reached Britain later than most other kinds of tree, but also use it, along with other types, to find patterns of temporary settlement by people who may have left few other traces for us to find. A decline in the pollen of trees like lime and oak, followed by the gradual appearance of birch and hazel – trees that colonise open ground rather than form part of a closed forest – is a sign of humans who have cleared areas of woodland, used the land for a while, and then moved on.

If bees are to make honey from lime, plenty of trees are needed. It has even been suggested, since lime is unique amongst native trees in flowering so late, that they could be planted on farmland as a late-season nectar crop. A few isolated specimens, considered Howes, were of little use. That is exactly the situation here – one tall specimen in a nearby garden, one planted to top the dome of a hill fort, and several that have established themselves on the edge of a woodland fragment not far from the brassica field where the bees are foraging now. Perhaps they visit the limes for nectar on the way, but no pollen appears in the traps.

The other late tree is no more common, but this time bees collect its pollen with enthusiasm. Sweet chestnut *(Castanea sativa)* is native to southern Europe and parts of Asia, and seems to have been brought here by the Romans. It coppices well, and in some parts of Britain has been widely planted. There is a small grove of mature trees within half a mile of the apiary, and two trees only yards away. These are smaller, but of flowering age, and the long yellow catkins, which have a few female flowers at the base, but are mostly male, produce a lot of pollen. Sweet chestnut pollen is brought to the hive throughout July, and may all come from these two trees. The grains are small, a little more oval than round, and the pollen loads a dull yellow. We will look at differences in pollen quality later on, but *Castanea* pollen has been shown in several studies to be particularly nutritious. One of these examined the effect of different pollen diets on the growth of the hypopharyngeal gland in nurse bees, the gland which produces the jelly fed to larvae as they develop. Chestnut pollen was among those that resulted in the largest glands, so the bees are quite right to collect it as avidly as they do.

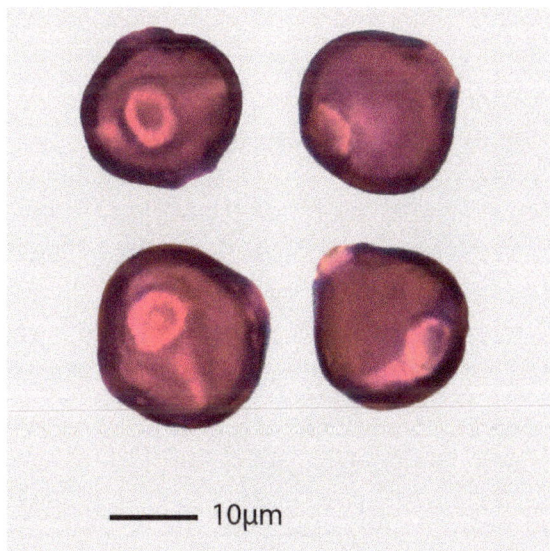

Figure 63: *Castanea sativa.* Family: Fagaceae (beeches and oaks).

It can be an important nectar plant too, and dark honey from chestnut forests is a well-known European type. There are fewer sources in Britain, but a survey of Worcestershire honey samples in 2016 found that one of the differences from Deans' results in the 1950s was that sweet chestnut had become more heavily used. Its pollen is a common type in honey. Pollen grains sucked into the honey stomach when the bee collects nectar are usually removed by a filtering process between the honey stomach and the gut. Depending on the type of pollen involved, and how long the process has to work while the bee flies from the foraging site back to the hive, much of the pollen can be eliminated. The filtration removes fungal spores and debris too, and no doubt helps reduce contamination of the honey store. American researchers in the 1940s examined the process by feeding bees with syrup containing a known concentration of pollen grains, and found that the level of pollen after filtration was only a 30th of the starting point. The efficiency varies between pollen types, though, and honey samples from caged bees feeding on forget-me-nots, which produce large amounts of tiny pollen, were once shown to contain 10-20 million grains per gram. Chestnut is prolific too, and the small size and perhaps the shape of the grains means that they often find their way into the honey jar.

The exposed position of the catkins, and the amount of pollen produced, suggest wind pollination. The grains themselves, though, are sticky, and clump together in the way of pollen carried by insects. Recent observations have confirmed that bees are regular visitors to chestnut flowers, and have no difficulty in gathering the pollen. Chestnut seems to be a plant which uses both methods: one idea is that wind transport is best in dry weather, and insects when there is rain. Its pollen can be carried somehow, whatever the summer is like.

So when bees come to the hive with their loads of chestnut pollen, I can be fairly certain of where they have been. Occasionally during the year there are pollens like this, where there are only one or two places where they can be found. The wood anemones whose pollen was collected in the first half of April are an example: there are one or two small clumps in the woodland around the apiary, but the only place nearby where they grow in numbers is on the slopes of a well-preserved section of Offa's Dyke, a hundred metres or so from the hives. For more widespread plants, pollen can only tell us the kind of habitat that the bees have used, rather than exactly where they have been. But even if we only know this, then gradually we can map out the foraging landscape in which they work.

This is a habitat map centred on the apiary, marked here by a red dot. The circle around the hives has a radius of 2 kilometres, and we will see the significance of that later on.

Habitats within 2km of the apiary.

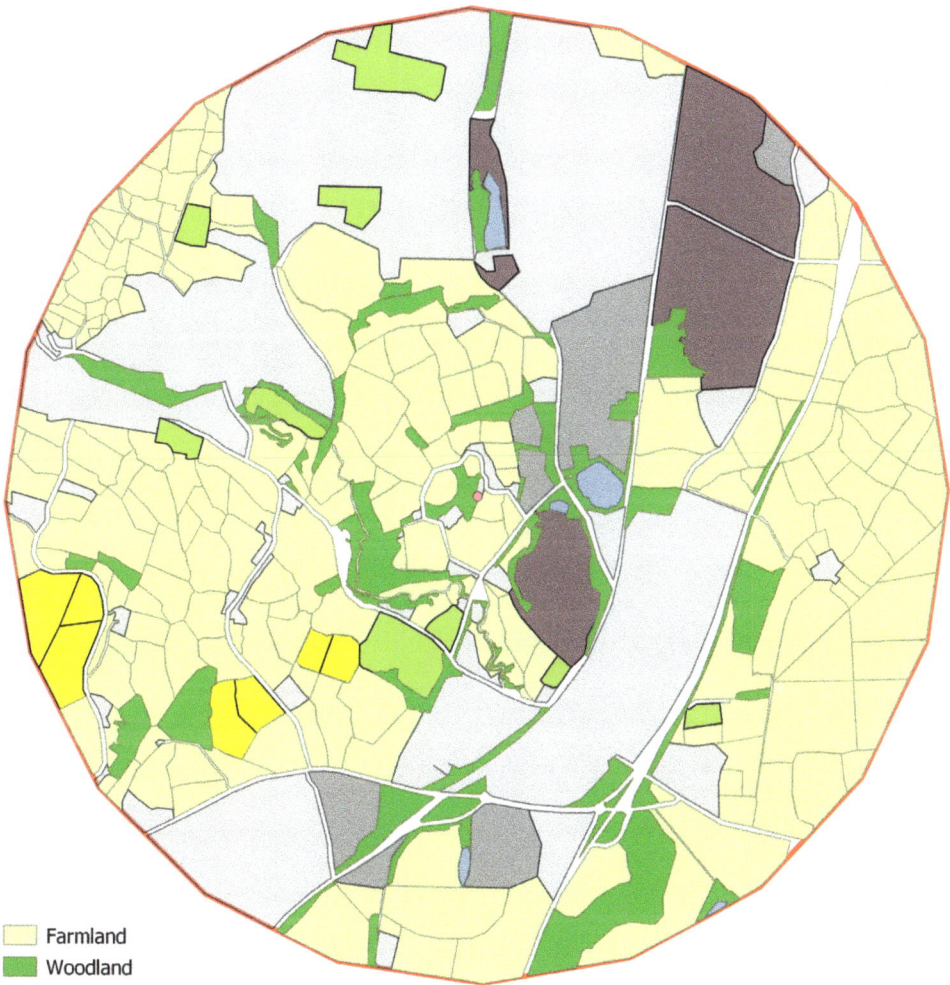

Farmland
Woodland
Brownfield site
Housing
Industrial estates
Amenity areas
Solar panels
Water

Figure 64: Habitats within 2km of the apiary.

The landscape shows the variety that you might expect from a place where there are traces of human settlement and industry over at least two thousand years. It is still changing: some fields are filled now with solar panels which have appeared in the last few years. There most of the clover has gone, but strips of wild flowers flourish around the edges and between the rows. The bees will have to continue to adapt.

The patches of woodland on the map are where they have mostly found pollen so far. Then there are the field boundaries, almost all of them hedges, with hawthorn the most common shrub, and those were important in May. They have found hogweed on road verges, and foraged on some plants that only grow in gardens. Perhaps a mixed landscape like this is the best kind, with different habitats providing resources at different times, and between them all providing food for insects all through the year. Mixed as it is, though, it is typical of most of the country in that the largest component is farmland. Even a country as urbanised as Britain is still about 70% agricultural land, and it is what the bees find there that is most important at this time of year. About a kilometre west of the hives are two cereal fields, unusually large for this area. Like most modern cereal fields, more effective weed control and seed cleaning methods mean that there are very few sources of nectar or pollen here. The tithe map shows that the same two large fields were there in the mid-nineteenth century, and that they were used for arable crops then too. The same map tells us that unless fields were swallowed up by industrial development in the rest of the century, or housing in the century that followed, their boundaries have changed very little. The main difference is in how they are used. When the map was drawn, the country relied heavily on home-grown grains, and lower yields required far more arable land than we have now. The habitat map includes some oilseed rape and cereals, and later in the year maize fields on the edge of the 2 km circle provide the other main annual crop, but most of the open ground here is grass. Since that makes up so much of the foraging landscape, it is worth mapping it out in more detail.

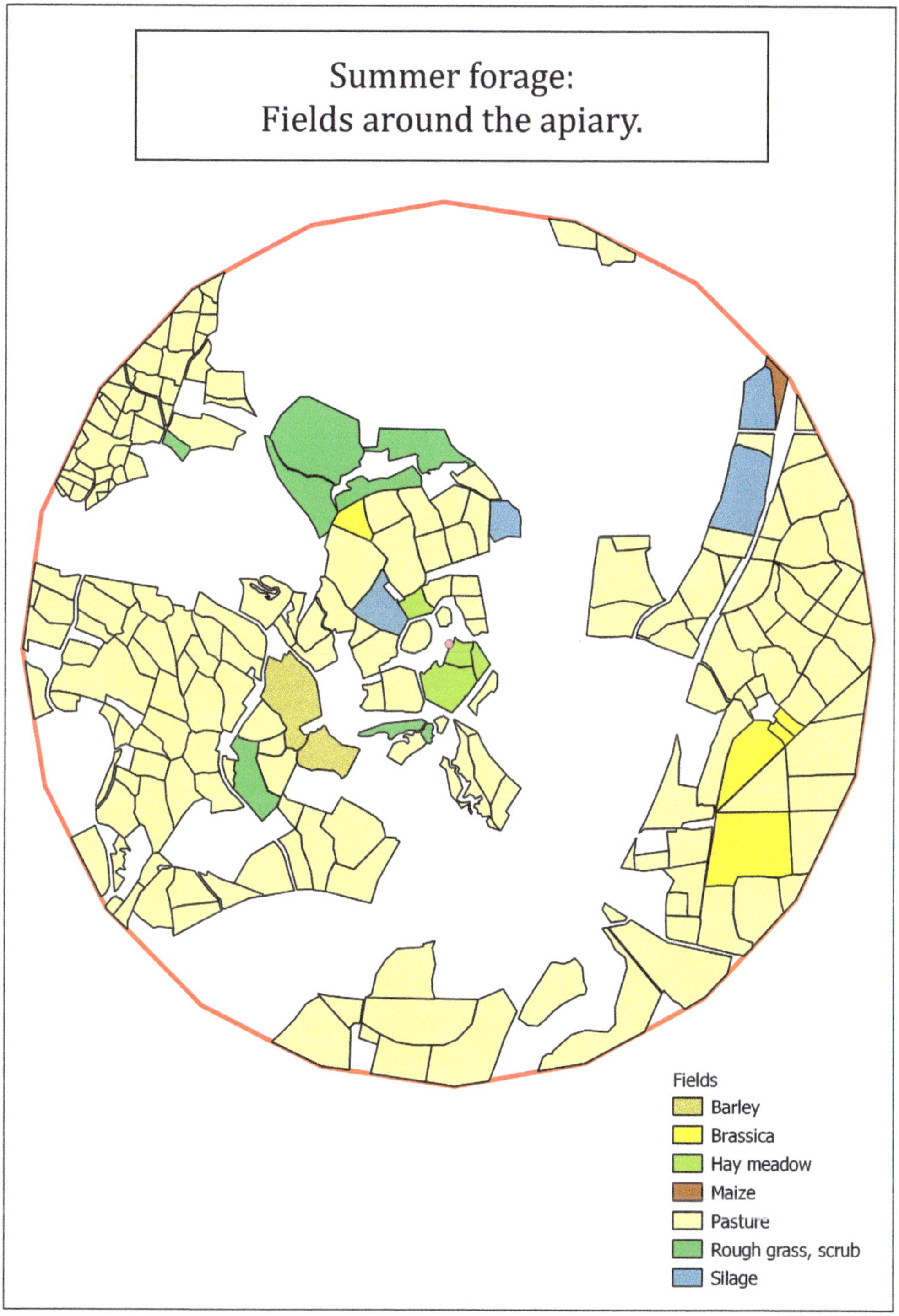

Figure 65: Fields around the apiary.

The fields lumped together as 'pasture' are not all the same. Some are grass and clover, with the uniform appearance of a carefully-managed crop. Most have varying amounts of other flowers: dandelions and their relatives, thistles or knapweed. Some are cropped close by sheep, and others trampled by cattle or horses. But it gives an idea of how widespread this type of habitat is in the pollen landscape here.

The fields which are sown and cut a few times a year for silage are almost pure crops of ryegrass. Once their hedges have stopped flowering, these fields have very little to offer, and even the hedges here seem to have fewer weeds than those along the smaller roads, where there can be a steady supply of bindweed flowers, either *Calystegia sepium* or *C. silvatica.* Bees collect whitish pollen from these, but white or white-grey pellets at this time of year can be from a number of plants. Under the microscope, though, *Calystegia* pollen is unmistakable.

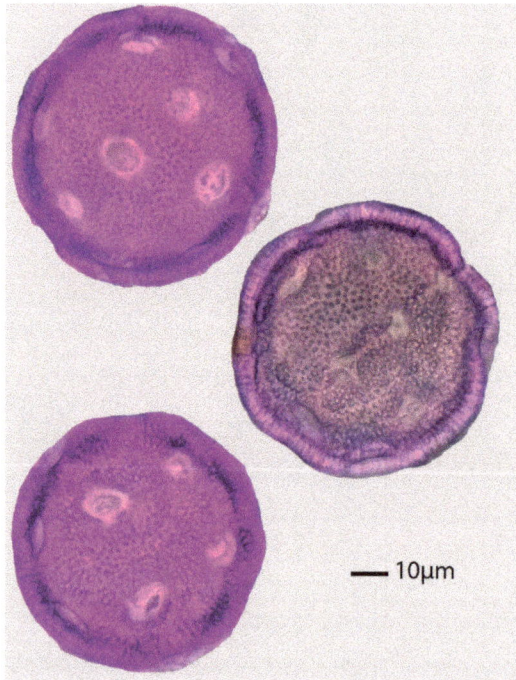

Figure 66: *Calystegia.* Family: Convolvulaceae (bindweeds)

The grains are unusually large, and the outer coat much thicker than in most types. The surface, interrupted by large pores, is dotted with outgrowths that are too short to be considered spines. They are usually compared instead with warts, and the surface does look something like the skin of a toad.

Where the grass is more permanent, there are usually patches of white clover, and both field and spear thistles. The thistles are two species of *Cirsium,* with pollen grains that this time do bear obvious spines, and which also turn up in pellets of an off-white shade.

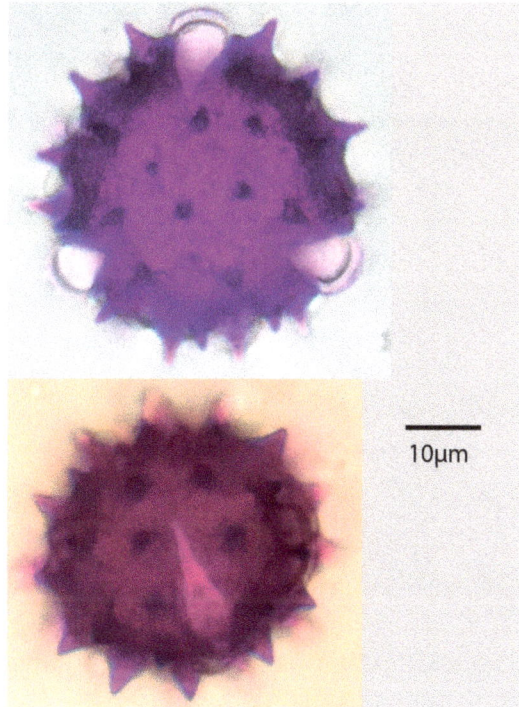

Figure 67: *Cirsium.* Family: Asteraceae (daisies)

Thistle flowers, like others in the family, are made up of a large number of individual florets, each of which can be a source of nectar or pollen. This may encourage bees to stay for some time on a single flower head, which in turn, it has been suggested, may leave them open to pollen theft. This behaviour, where honey bee workers steal pollen from bumble bees or solitary bees, has only rarely been observed, and in each case took place on flowers of this type. Another factor involved was the docility of the victim: some solitary bees flew off quickly when a honey bee approached, but others stayed to be robbed, and bumblebee victims rarely responded at all. Perhaps this kind of theft is no easier than gathering pollen of your own though, because, unlike large scale raids on the nectar stores of other honey bees, it hardly ever seems to occur.

July 2019, Week 2

Predominant		
Secondary	*Brassica*	
	Trifolium	Clover
	Castanea sativa	Sweet chestnut
	Filipendula ulmaria	Meadowsweet
	Rubus	
Important Minor	*Cirsium*	Thistle
	Centaurea nigra	Black knapweed
	Melilotus	Melilot
Minor	*Hypericum*	St John's wort
	Chamaenerion angustifolium	Rosebay willowherb
	Rumex	Dock
	Leucanthemum vulgare	Ox-eye daisy
	Aegopodium podagraria	Ground elder
		Fungal spores

The rough grass areas on the map are hardly managed at all. Places like these tend not to last. They may be developed for housing or supermarkets, or even if they are left alone will usually be invaded by scrub and then trees, and so lose their late-summer flowers. Here, though, some have survived for a long time and, along with the clover in the grazed fields and the more open parts of the brownfield sites, are likely to be the most valuable foraging areas in July and August. One of them, a damp field beside the river, has dense stands of meadowsweet, *Filipendula ulmaria,* and its pollen is in some years collected over several weeks. The loads are often a characteristic lime green, but sometimes they are more yellow and resemble the chestnut pollen with which meadowsweet overlaps for a while. The two types are also similar in size, and are both amongst the smallest grains that honey bees collect. But meadowsweet grains are round rather than cylindrical, and have shorter furrows and therefore a larger polar field.

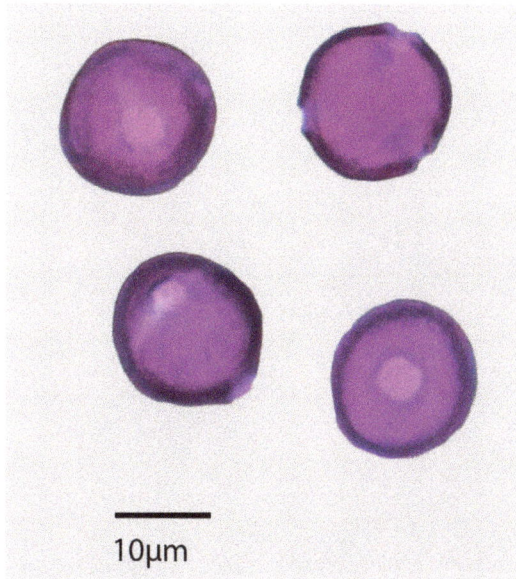

Figure 68: *Filipendula ulmaria.* Family: Rosaceae.

Moving up from the river on to higher ground, there are areas turned over for coal and then abandoned. Small spoil heaps and the craters of old shafts are mostly covered by grass, but there are brambles, and thistles and above all swathes of black knapweed, *Centaurea nigra.* All kinds of insects come to its flowers, and later in the year groups of goldfinches swirl and chatter around its seeds. It resembles its thistle relatives in pollen colour and is another of the white-grey pellets – pale pinkish grey was Mary Percival's description - in the samples in July. The microscope shows a surface of bumps, though, rather than spines, and it is easy to detect. Synge recorded it as a less-important pollen source, but it was one of the main types gathered by Percival's bees in July, and here the bees always collect it over many weeks. As we saw in chapter 2, it has three furrows, with a large, clear pore in the centre of each.

Knapweed and brambles compete effectively with grasses, and can establish themselves even amongst the tall tussocks of grass that cover some of this unmanaged land. Other flowers do better on different sites, where grasses are not quite so dominant. To the east of the apiary are places where clay was extracted. Some were abandoned more recently than the old coal workings, and in some the soil was so disturbed that the vegetation is slow to return. Two plants are particularly useful as pollen sources here. The first, melilot, has the word root for honey in its name, and is a well-known bee plant wherever it grows. Howes notes that it was even for a time farmed in the USA for honey crops. White melilot was the species grown there, but the yellow-flowered type, *Melilotus officinalis,* is the more common in this area, and does well on the sort of thin, nutrient-poor soils that industry has left behind. It is in the same family as clover, but the pollen loads are pale orange where the clovers are a darker brown. There is more of a resemblance under the microscope, where both are triangular if viewed from the pole, but form a cylinder when seen from the side. In each case, too, there is often a line of granules along the furrow.

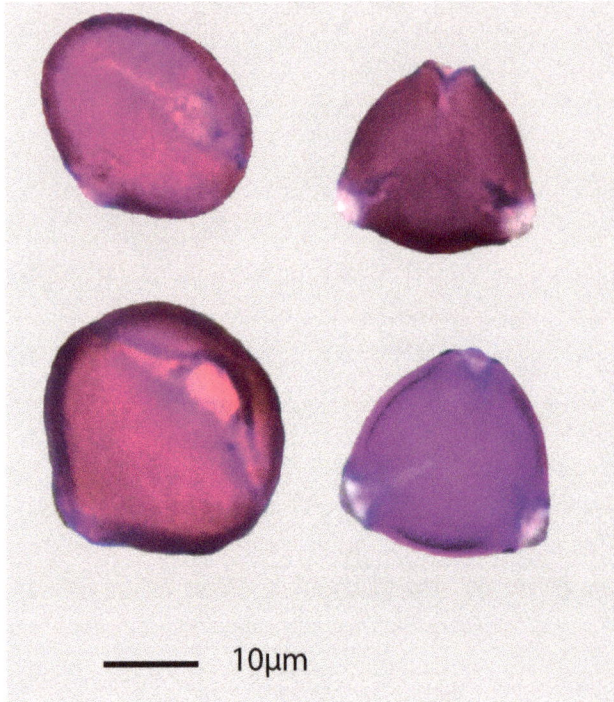

Figure 69: *Melilotus officinalis.* Family: Fabaceae (peas and beans)

The other pollen plant in this kind of habitat is probably the species most associated with disturbed ground: rosebay willowherb. Well-known for colonising bomb sites in the Second World War, it will appear on almost any open ground. It flourishes when conifer plantations are cleared, for example, or when industrial land is abandoned. There are temporary thickets in the woodland behind the apiary here, where storms have opened a clearing in the trees, and one of its names, fireweed, refers to its rapid occupation of land cleared by fire. Because it is mostly a plant of temporary spaces, its populations can vary wildly, but there can be few places where bees have no access to it at all. The beekeeper will always know, because the pollen is an unusual colour: a blue-black that is unmistakable at this time of year. Microscopically, the grains are also different. Not in the basic structure, which is a familiar three-cornered, three-pore type, but in their unusually large size, and the presence of long threads which can bind the grains together to form clumps on the slide. These are the viscin threads, found only in a few families of flowering plants, including the Onagraceae to which the willowherbs belong. Evening primrose is another member, and its enormous pale yellow grains are also bound with viscin threads. Viscin is the name given to the material which surrounds mistletoe berries, which is so sticky that years ago it was one of the components of bird lime smeared on to branches to catch birds. It seems likely, though, that the viscin threads of pollen are not sticky at all, and simply tangle with each other and with the bristles of insects like bees. The result is that the pollinator, if it reaches the stigma of another plant, is likely to deposit a clump of pollen grains together, and it was proposed about 40 years ago that this might be a way for the flower to improve what is called its pollination efficiency. A flower of perfect efficiency would fertilise each egg with a single pollen grain. Pollen, expensive to produce, would never be wasted. At the other extreme, wind pollination is so inefficient that huge numbers of grains have to be made, and most of them are lost. As we saw earlier, using animals for targeted delivery is a way of increasing that efficiency. The next step might be to improve the service by having each delivery fertilise many eggs at once. So rosebay willowherb, like others in its family, has large numbers of ovules in each flower, and delivers its pollen in tangled clumps to fertilise them all. Greater efficiency means that less pollen has to be made – but there is still enough, and over a long enough season, to make this a useful pollen plant for bees.

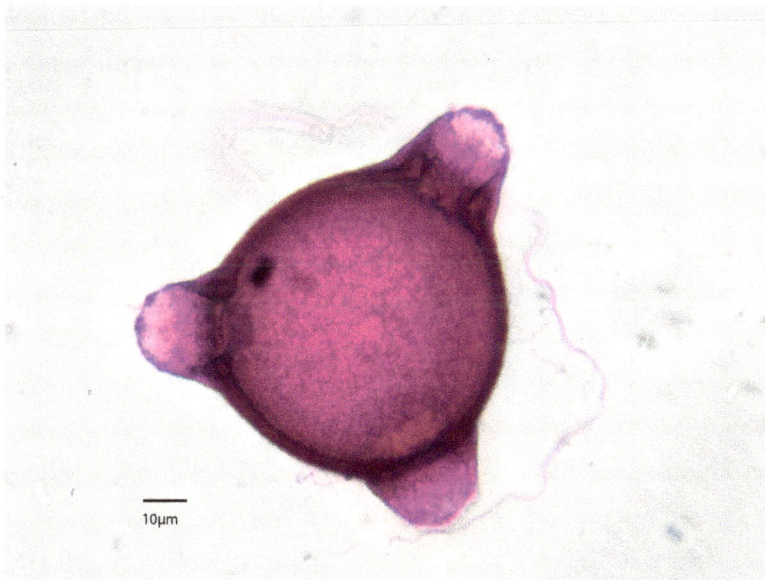

Figure 70: *Chamaenerion angustifolium.* Family: Onagraceae (wilowherbs, evening primroses)

Another of the families where the pollen grains grow viscin threads is the one to which the heathers and the rhododendrons belong, the Ericaceae. When we looked at their pollen earlier, we saw that the grains were already clumped into groups of four. In species which have the threads, each clump of four becomes attached to several others, to produce what a recent Chinese study described as pollen thread tangles. The authors of this investigated another possible angle of pollination efficiency. What if you are a flower whose pollinators are relatively scarce, or don't visit very often ? In such cases, loading them with big clumps of pollen would be the best thing to do. The results from China supported this, showing that species with fewer visits from pollinators produced larger pollen tangles. The pollinator makes fewer journeys, but carries more pollen each time.

There are two other kinds of grassland on the map, and on a national scale one has decreased dramatically and one become more common. Hay meadows are now an unusual habitat and the one near to the apiary is, as I have described, a recent and maybe temporary example. Their advantages for insects are well-known: they are cut late in the year, allowing the plants to flower and so provide food. From the narrow perspective of pollen for honey bees, this particular meadow offers nothing they can't find elsewhere, but some of the pollen of brambles, knapweed, thistle and above all clover probably comes from here.

The more common habitat is the road verge. The smaller lanes on the map run between hedges but larger roads usually have a grassy verge, and in total these can make up a significant grassland habitat. The plant conservation charity, Plantlife, lists more than 800 plant species that can be found there across the UK. Sometimes the sort of management that favours the most diversity – fewer cuts and a less tidy landscape – can also save money, and many local authorities have begun to consider verges from a conservation point of view. Some plants that would once have been common in hay meadows find a home there, including the ox-eye daisy, *Leucanthemum vulgare*. Its pollen is a bright orange and stands out in the samples from the duller shade of melilot. The daisy family is a successful one, and the spiny grains could come from several species. Many, though, are ignored by bees, others produce pollen of a different size, and the large numbers of *Leucanthemum* that are flowering now makes it the most likely source.

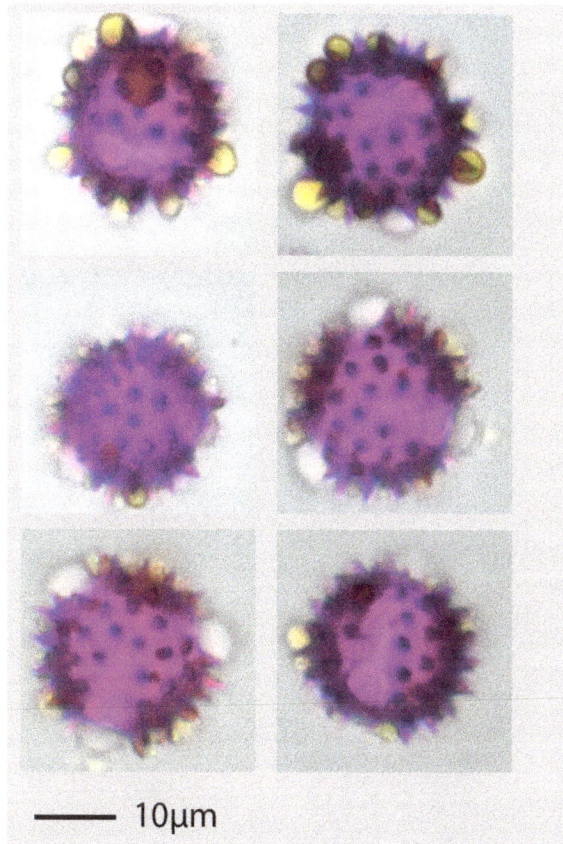

Figure 71: *Leucanthemum vulgare.* Family: Asteraceae (daisies)

July 2019, Week 3

Predominant		
Secondary	*Trifolium*	Clover
	Castanea sativa	Sweet chestnut
	Melilotus	Melilot
Important Minor	*Filipendula ulmaria*	Meadowsweet
	Rubus	
	Chamaenerion angustifolium	Rosebay willowherb
	Brassica	
Minor	*Cirsium*	Thistle
	Hypericum	St John's wort
	Taraxacum type	
	Fuchsia	
	Calystegia sepium	Hedge bindweed
	Hydrangea	

Another of the orange pollens could also come from the roadsides, where many of the yellow-flowered relatives of the dandelion grow. The hawkbits and hawkweeds, cat's ears and sow thistles are taxonomically like brambles – a thicket of types that only a specialist would try to separate. From a pollen point of view, the grains look much like those of dandelion. Different from any other group, but almost identical to each other, the best we can do with this pollen is to be like the casual botanists who lump them together as yellow composites (after the Compositae family to which they belong) and list them in the table as being like dandelions – *Taraxacum* type. When investigators in New Zealand did the same thing, they also found an expert to identify the plants from which the pollen might come, so that they had an idea of how many species they were lumping together in the group. As well as dandelion itself, the survey turned up five types of hawkweed, smooth hawksbeard and cat's ear. None of these plants occurs naturally in New Zealand, so although they can be harmful invasives and quite common sometimes, there are relatively few types compared with the choice that bees in Britain have. The table entry here may cover many kinds.

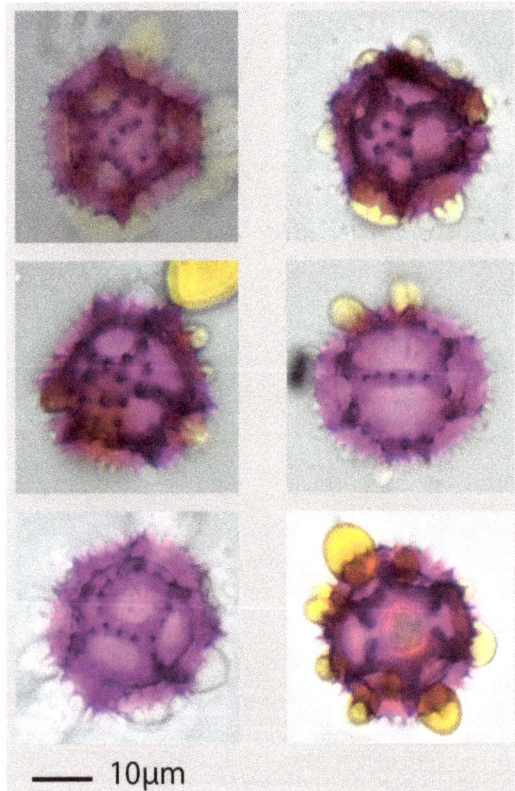

Figure 72: *Taraxacum* type. Family: Asteraceae (daisies)

Lumped together, they are as widespread as dandelions too. The yellow flowers brighten road verges, but are common wherever there is some grazing or cutting to hold the grasses back. They may not be among the bees' favourites, though: the New Zealand work suggested that the pollen was only collected when there were few alternatives, and an Irish study came to the same conclusion, with the dandelion relatives abandoned once brambles came into bloom.

There are no urban centres on the habitat map here, but the main areas of settlement, mostly former industrial villages with small houses and small gardens, still cover a fair proportion of the land inside the 2km circle. Almost fifty years ago Jennifer Owen began one of the first ecological studies in a suburban garden. She recorded the insects in her Leicestershire garden for thirty years and showed that a habitat usually ignored for being 'unnatural' could support a surprisingly rich variety of life. It was an important finding because, as with road verges, the total area of gardens adds up to a significant amount of land. Since bees are often kept these days in the centre of the largest cities, where gardens and other urban planting is probably all they can reach, they can clearly find food in such places, and their ability to forage on all kinds of plants has served them well. Garden pollen may rarely be the main kind collected, although this year we have seen exceptions for California lilac and Welsh poppies over a couple of weeks in the June gap, but it makes a contribution most of the time. Often, it may be the kind of small amounts that a pollen trap can miss, but in the latest table we have the distinctive lemon of *Fuchsia* pollen, and the tiny grains of *Hydrangea* turn up amongst the white-grey loads. Another of the grains with the common arrangement of three furrows, hydrangeas also have plenty with four, as in this middle example on the lower row:

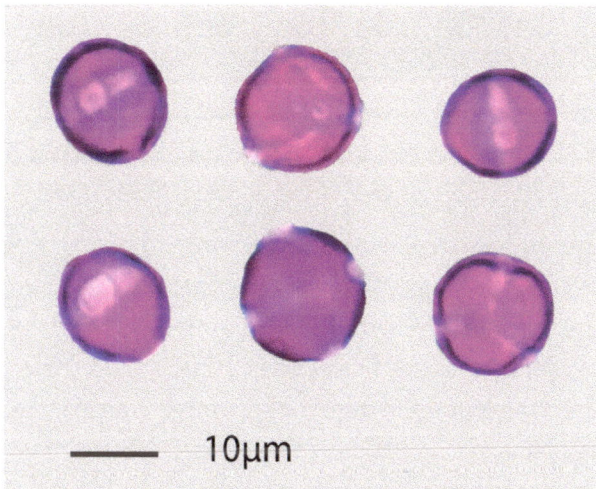

Figure 73: *Hydrangea*. Family: Hydrangaceae

Hydrangeas are a staple in many gardens, and a large bush in flower looks as if it should have as much to offer bees as a blossoming fruit tree. Hydrangea flowers, though, come in two kinds. Those with nectar and pollen are small and inconspicuous. They are surrounded by large, colourful flowers which attract pollinators, but are sterile, and so have no pollen to give. It is another way, perhaps, in which plants advertise rewards without having to give too many of those rewards away. We have then bred hydrangeas for gardens with more of the showy flowers, and fewer of the small ones, and in the end the whole bush of the typical mop-head type may have little to offer bees. Other kinds, though, do: the climbing hydrangea has small, fertile flowers, and the name of the lacecap types reflects a structure where a ring of the sterile flowers surrounds a flat cap of the smaller type. On these kinds, bees collect pollen into the start of August.

Fuchsias produce pollen at the other end of the size scale from hydrangea, with each grain about four times the size. They are in the same family as rosebay willowherb, and the pollen is of a similar type, both in shape and in having viscin threads to bind the grains to each other.

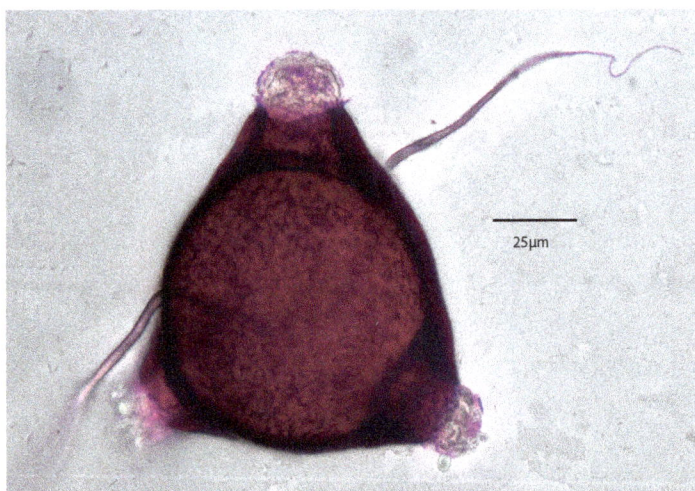

Figure 74: *Fuchsia.* Family: Onagraceae (willowherbs and evening primroses)

As much as any plant, perhaps, fuchsias show how flexible bees are in where they find their food, and how they can thrive in gardens almost regardless of the kinds of plants that each gardener grows. The original home of most kinds of fuchsia is south and central America, from where they were brought to Britain at the end of the 18th century. It is the other side of the world from where honey bees evolved, and the shape and colour of the flowers is typical of a plant pollinated by birds rather than insects, but bees can use them. In parts of Ireland, and in the south-west of England, the climate suits them so well that they can establish themselves outside gardens, and are even used sometimes as a hedging plant. Then they can become a significant bee resource, especially for nectar, because a flower ready to feed birds has plenty of nectar to give. Here in north Wales, perhaps, the winters are not mild enough for this to happen, but the rainy climate suits them, and individual bushes are a common sight. Bees climb up into the flowers on mild days well into the autumn, though they have stopped collecting the pollen, and visit them only for nectar then. There are even some wet summer days when fuchsia pollen makes up most of what the bees collect. This may be partly because there are several shrubs close to the hives, and the bees are not going far in the rain, but also because the hanging flower tubes with their outspread sepals are like umbrellas above the pollen, and keep it available even on rainy days. The pollen is usually distinct in colour, being much more of a lemon shade than most yellow types, and it also has an powdery appearance, reminiscent of the dry grains of a wind-pollinated plant.

July 2019, Week 4

Predominant		
Secondary	*Trifolium*	Clover
	Rubus	
Important Minor	*Filipendula ulmaria*	Meadowsweet
	Castanea sativa	Sweet chestnut
	Melilotus	Melilot
Minor	*Chamaenerion angustifolium*	Rosebay willowherb
	Fuchsia	
	Clematis vitalba	Old man's beard
	Calystegia sepium	Hedge bindweed
	Cirsium	Thistle
	Hypericum	St John's wort
	Taraxacum type	
	Hydrangea	
	Centaurea nigra	Black knapweed
	Ligustrum	Privet

Clematis pollen collected earlier in the year came from gardens too, but the type that appears now, yet another mixed in with the various shades of grey and white, is from *Clematis vitalba,* the old man's beard of roadside hedges. Although towards the northern end of what seems to be its natural range, it is common here, and a source of nectar as well as pollen. It is in the buttercup family, like the celandines and wood anemones that were visited at the start of the year, and its pollen is similar to those.

Honey bees can gather food from all the habitats shown on the map around the apiary. From abandoned industrial sites to gardens to fields on modern, efficient farms, on any of them they will find plants that they can use. But few of these types of habitat on their own would be enough, and in some places we have made landscapes, less variable, where even this most adaptable of insects can struggle. And if honey bees find it hard, then we can be sure that many other kinds of pollinators will suffer more. Some will have more specialised needs, some will not be able to travel far enough to find what they need.

All lack one of the advantages that social life offers honey bees: that a colony of foragers can search a wide area to find food resources even when these are scattered over farms and towns. How wide is what we will look at next.

8. How far do they go ? August 2019.

Hedges with clematis grow not far from the hives, and there are fuchsia bushes closer still. Ecologists have plenty of evidence to support the common sense idea that, other things being equal, animals should gather their food in a way that brings in the most energy for the least effort. With cats, that seems to involve sleeping for most of their lives; for bees, which don't have that option, at least finding food close to the hive would reduce the fuel expended when they fly. How far they really go, though, is still the subject of research.

François Huber, whom we last saw investigating why bees collect pollen, thought that knowing how far bees foraged was useful, because it would help establish how many colonies an area could support. In his New Observations Upon Bees, published in 1792, with a second volume added in 1814, he concluded that there was still uncertainty about

> the greatest distance that bees ramble from their hives to gather their harvest. Different authors assert that they may wander several leagues from their home; but from the few observations that I have made, I believe this distance much exaggerated. It appeared to me that the radius of the circle they traverse does not exceed half a league [about 1½ miles]. Since they return to their hive with the greatest speed, when a cloud passes before the sun, it is probable that they do not go very far. Nature, which has inspired them with such terror for a storm and even for rain, evidently does not permit them to stray at distances which expose them too long to the injuries of the weather.

A modern ecologist might make more of the economics of flight than fear of the weather, but Huber's estimate of distance would turn out to be broadly correct. Even his fierce critic, Robert Huish, left it unchallenged. A year after the appearance of Huber's second volume, Huish published his own *Treatise on the Nature, Economy, and Practical Management of Bees.* A contemporary assessment has him as 'an obscure and an unscrupulous scribbler', and the *Dictionary of National Biography* tends to agree, but does acknowledge that bees seem to have been the one subject on which he could speak as an expert. He achieved nothing like the fame of Huber, however, which may explain his views in the 1844 version of his Treatise, where he described the discoveries of the Swiss naturalist as 'the result of fiction and delusion' and himself as standing alone in the defence of truth. Huish produced all kinds of books, from royal biographies to accounts of Arctic exploration, and novels described in the *Dictionary* as being 'of a very low type'. One title, *Fatherless Rosa; or, The dangers of the female life: expressly written as a companion to Fatherless Fanny*, may give a flavour of those. But he also expanded and updated his original bee manual through various versions, and invented the 'Huish hive', a kind of top-bar hive made from straw. It was still common practice at the time to harvest honey by killing the whole colony at the end of the season with sulphur fumes, and Huish's hive was a move towards the modern method of only removing some combs and allowing the colony to survive. And although his objections to some of Huber's opinions seem not be based on any investigations of his own, he did take an experimental approach to the matter of how far bees fly. Wondering whether it would be worth moving some hives closer to an area of heath three miles away, he left an observer at the hives and went himself to wait by the heather, armed with a pepperpot of flour. When he saw bees on the heath, he sprinkled them with flour, and some of these dusty bees then turned up at the hive. He also noted that the bees he saw foraging on heather on the Bass Rock in the Firth of Forth would have had to cross at least a mile of water to get there, and then have another mile to carry their nectar home.

Observations like this became the basis of what was known about how far bees will travel for food. The German 'bee Baron', August von Berlepsch, was another pioneer of the modern hive with its removable frames, and also happened to find himself in a position to investigate foraging distance. There are a number of different sub-species of honey bee, and although they hybridise freely, unmixed populations can sometimes look quite different from one another. Baron von Berlepsch was the only beekeeper in his locality to keep the Italian race, and was able to identify them by their colour. By searching the plants around his hives, he found his bees usually within two or three kilometres of home, but in one year as far as 7 km away.

A circle with a radius of several kilometres forms a huge area in which to search for small insects. Huish's method of examining the bees when they come back to the hive is much easier, as long as you can have some idea of where they have been. Huber had already suggested as much: 'For a conclusive experiment on this subject, it would be necessary to make it in vast arid or sandy plains, separated by a known distance from a blooming country'. Although Huber himself had no such landscape, John E Eckert in the United States did. Eckert was born in Ohio in 1895, and spent most of his working life in the University of California. But in 1927, he was an Associate Apiculturist with the United States Department of Agriculture, based in Laramie, Wyoming, with all the open prairie he needed to try out Huber's idea. He found two irrigated areas with 17 miles of badlands between them, placed hives at different points in the badlands, and recorded the pollen that the bees brought back. Even bees starting from the centre of the badlands returned with pollen from the crops in the watered zones, and so must have made a round trip of at least 17 miles, or 27 km, each time.

von Frisch studied the question of distance too, as part of his investigations into the bee dance. Like Huber, he had the problem of the more cramped European landscape, and this he overcame by carrying out his experiments in autumn, when there was little natural food for the bees, and they could be trained to use artificial feeders at a known distance from the hive. As Huish had done, von Frisch marked the bees which visited his feeders, although with paint now rather than flour, and found that they would come to a feeder as far as 12 km from the hive. Considered as a round trip of 24 km, this agrees quite well with Eckert's results, and yet in both cases the question answered is not so much how far *do* bees fly, as how far they *can.* Rather than storing honey, the colonies that Eckert placed more than a few miles from the crops tended to lose weight as they used up their stores of food – in other words, the colony was using more food than it was able to collect. Only desperation was taking them so far. Honey bees are able to travel great distances, considering their size, but long-term survival requires that they find food much closer than this. The real question about distance, then, is where bees go under normal conditions. Eckert tried to answer this too, by placing bees in an agricultural area in Colorado. Like von Berlepsch, he made use of the colour differences between different types of bees, this time setting out colonies of Caucasian bees which he could distinguish from the Italian bees that the local beekeepers used. And then he and two helpers walked through fields of clover in the area around the hive, and looked for his bees.

Even in this kind of landscape, Eckert found that the bees sometimes flew about 4.5 km from the hive. Strangely, they seemed to have a direction that they preferred – what Eckert described as a 'flight lane' – and would visit fields to the north west of the hive which were further away than areas in other directions which seemed just as attractive – to the human eye, at least.

In Britain, with summer weather much less stable than Eckert's bees had enjoyed, the advice in a 1949 leaflet from the National Agricultural Advisory Service was that bees needed to have adequate pollen and nectar supplies within about 2 miles or 3.2 km of the hive. A year earlier, at the height of his successful career as a commercial beekeeper, R O B Manley suggested a radius of about 2.5 km. His concern, as someone who made his living from honey, was honey production, and how close bees had to be to their food source in order for a surplus to be made. Since a surplus is what allows survival over the winter, distance that allow it are what we need to find. With this in mind, experiments were done in the summers of 1949 and 1950 near the same Rothamsted research station where Synge was collecting pollen, measuring the weight changes of hives placed at different distances from crop fields, or from sources such as lime or heather. The hives were never more than 1.2 km from the crops, but even these flight distances had an obvious effect on their foraging, and in the poor summer of 1950 the colonies were unable to gain weight at all. It began to seem as though the useful foraging distance was quite a bit smaller than the maximum, although any results have to be qualified by saying how much things are affected by the weather, by the landscape type, and perhaps also by how much competition there is for food.

August 2019, Week 1

Predominant	Trifolium	
Secondary	*Rubus*	
Important Minor	*Filipendula ulmaria*	Meadowsweet
	Chamaenerion angustifolium	Rosebay willowherb
	Leucanthemum vulgare	Ox-eye daisy
Minor	*Fuchsia*	
	Calystegia sepium	Hedge bindweed
	Cirsium	Thistle
	Melilotus	Melilot
	Chenopodium	
	Echinops	Globe thistle
	Zea mays	Maize
	Castanea sativa	Sweet chestnut
	Plantago	Plantain
	Taraxacum type	
	Hydrangea	
	Centaurea nigra	Black knapweed
	Graminae	Grass

Two distinctive kinds of pollen turn up as something new in the first August sample. *Chenopodium* pollen is collected in dull brown loads that may have come from clumps of fat hen *(Chenopodium album)* that are flowering on waste ground and the edge of some of the grass fields. It has pores across the surface, a little like plantain pollen, but these grains are a little smaller and there are many more of the pores.

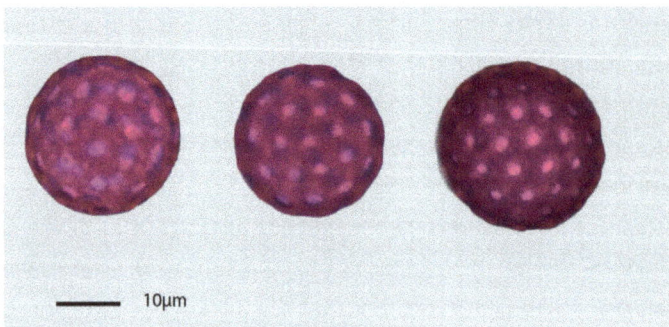

—— 10µm

Figure 75: *Chenopodium album.* Family: Amaranthaceae

Enormous white-grey grains, meanwhile, are from a garden plant, one of the kinds of globe thistle, *Echinops.* They have the spines of pollen from the daisy family, but over such a large surface these appear short and blunt.

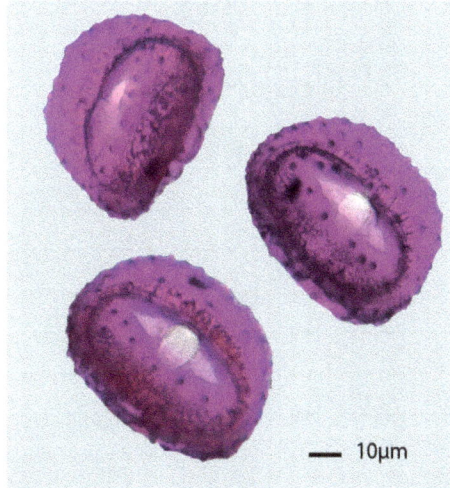

Figure 76: *Echinops.* Family: Asteraceae (daisies)

Chenopodium is too widespread and *Echinops* too scattered to tell us much about foraging distance, but other plants in the August sample can be of more help. Pollen from crops, like those used in Eckert's study and the Rothamsted work, are useful even without marking bees. Other than the odd stray, crop plants are confined to fields, and if we know where those fields are, they can suggest how far bees have flown. At the start of August 2019, maize pollen appeared in the collection. The pollen of the grass family, to which maize belongs, all tends to look alike, as we have seen, but many have finished flowering by now. In addition, pollen from maize is more of an amber colour than the typical yellow of grasses, and although the shape is that of any grass, the grains are unusually large. When we see these, we can be fairly sure how far the bees have flown.

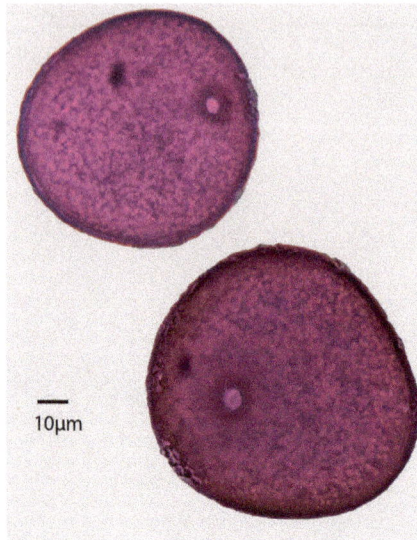

Figure 77: *Zea mays.* Family: Gramineae

Maize has become a more common sight locally over the last few years, as it has in many parts of the UK, where we are approaching half a million acres of the crop. But around the apiary it is only grown on the lower ground to the east, and figure 65 shows that the bees have had to fly almost 2 km to find it. The same land is where most of the brassica crop is found too, much earlier in the year: another hint that the bees are happy to fly at least this far for food. But while brassica pollen is sometimes the main type collected, maize is never common. Perhaps there is something about brassica pollen that makes it worth travelling a long way to collect, or perhaps they have fewer choices at that time of year and have to fly a long way to supply the hive. We can come back to the question of pollen preferences once we have looked at what we know now about how far foraging bees go.

A year after his field season in Colorado, Eckert moved to begin his new job in California. He studied bees for the rest of his working life, and in 1942 made the first estimate of how much pollen a whole colony might collect, suggesting between 44 and 125 pounds in a year. He would have marvelled at the discoveries of von Frisch once they became known. Having considered the problem himself, he might well have predicted how interpreting the bee dance would eventually throw new light on where and how far honey bees go. One of the first studies to do this was one we have looked at already: the work of Visscher and Seeley in the forest of New York State. They watched bee dances in an observation hive, over 36 days in the summer of 1980. The average distance signalled by the dances was 2.26 km, so something similar to the 2km radius circle marked out on my map. But there was a lot of variation around this: dances indicated anything between 50m and just over 10,000m. So another way of thinking about where the circle should be drawn is to consider what proportion of journeys it should include. If we want to know everywhere the bees can go, then the circle has to contain even the furthest destination. But then the area involved becomes so great that it is difficult to know in detail what vegetation it includes. Reducing the radius makes the plant survey more manageable, and still includes a good proportion of the foraging trips. So in the New York study, for example, a circle of a little more than 10km radius would be needed to contain every trip, but going down to 5km would still enclose 90% of them. Even a 1.6km circle would include more than half. And the distances here were far greater than earlier work had suggested, something which Visscher and Seeley put down to those having been carried out in agricultural areas. Their own results, they suggested, gave a better picture of what bees did in a natural setting. For my apiary, set in its mix of habitats, a 2km circle (about 12.5 square kilometres) is likely to give a good idea of where most of the bees go.

For all the advantages of using the dances to see how far bees have been, an obvious doubt is whether the dancing bees represent what the colony as a whole is doing. What if many of the bees are neither dancing nor following, but just doing their own thing, perhaps finding food much closer to the hive than the dances suggest ? Then the dance results would be misleading, but Visscher and Seeley argued that that wasn't the case. Some of the dancing bees were dancing about nectar, but others were signalling a pollen source, and they would carry that kind of pollen on their legs. So if the colony was mostly following dances, then the colour of the pollen brought in by the foragers ought to match the colour on the legs of the dancing bees. If most of the dancers carried the grey pollen of brambles, for example, then most of the returning foragers should have grey pollen too. If only a few bees danced with orange pollen on their legs, then only a few orange loads should be brought back. And that was what happened, suggesting that dance information is a reliable way to find out about the foraging behaviour of the hive. Gradually, too, it became a little easier to decode. Developments in video technology, and in computer storage and processing made it possible to record more dances than ever and then go through them frame by frame. These days, results are measured as the average of several dances mid-performance, for the greatest accuracy, and the time noted with radio-controlled clocks in order to match them with the position of the sun. There is one homely touch amidst all the technology: a vertical line against which the angle of the dance is measured is still sometimes created by nothing more than a washer hanging on a thread.

August 2019, Week 2

Predominant	*Trifolium*	
Secondary		
Important Minor	*Rubus*	
	Filipendula ulmaria	Meadowsweet
	Chamaenerion angustifolium	Rosebay willowherb
	Leucanthemum vulgare	Ox-eye daisy
	Centaurea nigra	Black knapweed
Minor	*Fuchsia*	
	Melilotus	Melilot
	Zea mays	Maize
	Plantago	Plantain
	Hypericum	

With some important exceptions, the tables from here will begin to show pollen types disappearing from the menu, without being replaced. There is no more sweet chestnut, no thistle pollen this week, and the meadowsweet is almost gone. Plantain pollen reappears though – it is flowering again in the nearest field after being cut, and may also be gathered from roadsides. White clover remains the main pollen source for the hives, as it will be for the rest of the month.

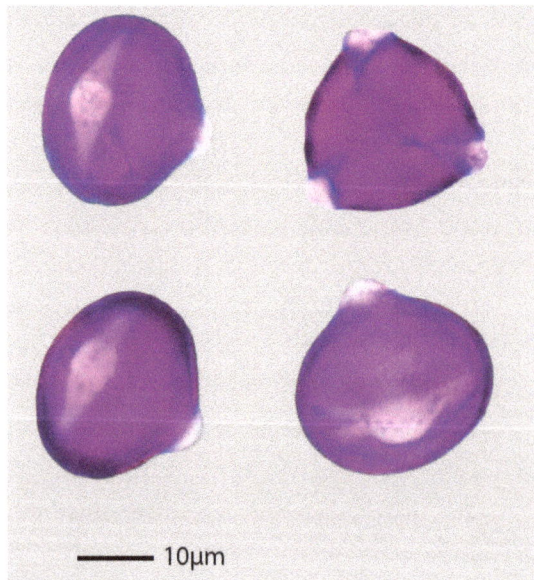

— 10µm

Figure 78: *Trifolium repens*. Family: Fabaceae (peas and beans)

The distance results from New York suggest that a good answer to the question of how far bees go is – it depends. Or put another way: as far as they must. And that will vary, partly with the type of landscape, as Visscher and Seeley said, and also with the season. There will be times in the year – exactly when might vary from place to place – when bees have to go further to get what they need. When the social insect research unit at Brighton looked at this in 2014, they found that the average distance travelled was only 493 metres in March and April, but just over 2.1km in July and August. Represented as a circle around the hive, the bees were searching an area of 0.8 km² in spring, but foraging over 15.2 km² in the summer months. The authors suggested that land management changes, like the steep decline in hay meadows that would once have provided summer flowers, might be one factor in this. Perhaps beekeeping folklore is out of date, and we need to forget about the June gap, and think of a late summer gap instead.

That may be true in Brighton, but what about elsewhere ? Decoding bee dances is difficult work, for all the help that technology gives, and we don't yet have many studies of this kind. One example, from Sheffield, was actually carried out several years before, and both supports the Brighton results and hints at how local conditions can affect them. Foraging distances here were again much greater in August than in May, but now the August distance was enormous: an average of 5.5 km, with 90% of the bees foraging beyond the 500m that was the spring average in the Brighton work. So almost all of the bees were having to travel a fair distance to find food – a summer gap indeed – but the great length of the average journey seems here to have been the result of a particular feature in the landscape. To the west of the city, heather moors bloomed in August, and offered a rich enough resource to draw so many bees so far. It is a dramatic example of the benefits of colony life. A social insect, with the ability to communicate where food is, can exploit a rich patch of flowers even a long way from the hive.

Lowland heath, as opposed to the grouse moors of upland districts, is another habitat which was greatly reduced during the last century. Its loss may be a further reason why the Brighton bees, living in a region where lowland heath was once much more common, had to travel so far in late summer. Heather moorland is still widespread across much of upland Britain, though, and there it often covers large blocks of ground with very few other types of flowering plant. This can result in a kind of honey unusual in this country, made almost entirely from the nectar of one or two species. Such monofloral honeys, incidentally, are certified on the basis of the pollen they contain. As we saw with chestnut pollen in the last chapter, many factors affect how much of each type ends up in the honey, and although new DNA sequencing techniques have made it easier to identify the grains, honey analysis is not straightforward.

Partly to make heather honey, and partly because heathers can sometimes be the main flower source available at this time of year, beekeepers often move their hives up to take advantage of moorland flowers. Young heather plants make more nectar than old bushes, and so honey bees may do best on moors managed for grouse, where areas of young heather are created by burning patches of ground on which new shoots then grow. Under the right conditions it makes a rich nectar source, and bees living amongst it rather than having to fly six kilometres from Sheffield can soon fill their stores. Like residents in a seaside town overrun with tourists, the local bumblebees must suddenly find themselves competing with thousands of interlopers. From a distance, as whole slopes come into flower and the colour of the landscape changes, it looks as though there is enough for all, but the weather at that height can easily keep honey bees confined to the hive, and Howes concluded that a good honey harvest might only come one year in seven. Even so, heather offers a late-season chance for bees to at least build up their own winter stores, and a commercial beekeeper like Manley still found it worth doing, sometimes in the 1940s loading more than 300 hives on to lorries and moving them to the moors. As well as the uncertain harvest, there were other drawbacks to this migration. He found his bees particularly bad tempered after their journey, nor did the change of scene seem to improve their mood, and he wrote that bees working heather were 'almost always unusually spiteful.' Even so, heather honey sells well, and a late summer move to moorland is often still part of the beekeeping year.

August 2019, Week 3

Predominant	*Trifolium*	
Secondary		
Important Minor	*Chamaenerion angustifolium*	Rosebay willowherb
	Leucanthemum vulgare	Ox-eye daisy
	Centaurea nigra	Black knapweed
Minor	*Rubus*	
	Filipendula ulmaria	Meadowsweet
	Fuchsia	
	Melilotus	Melilot
	Taraxacum type	
	Zea mays	Maize
	Calluna vulgaris	Heather
	Calystegia sepium	Hedge bindweed
	Graminae	Other grass
	Viola	
	Clematis vitalba	Old man's beard

As the bee flies, there is a hillside of managed moorland four kilometres from the apiary here. It is well within the range reported from Sheffield, but although heather-feeding bees collect pollen as well as nectar, in some years very little heather pollen is brought back to my hives. There are so few that the bees are probably not visiting the moorland at all, but finding heather plants in gardens or even some patches of waste ground nearer to home. The pellets are a very pale brown, or sometimes a muddy white, and the grains have the tetrad arrangement found in rhododendrons and winter heaths.

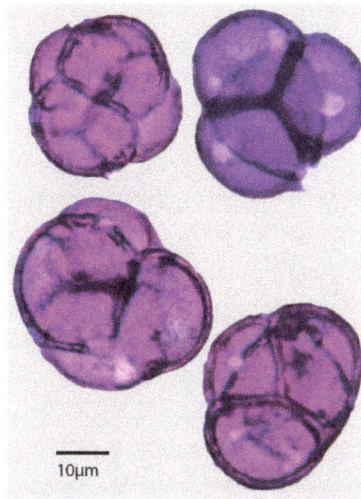

Figure 79: *Calluna vulgaris.* Family: Ericaceae (heaths)

When bees dance, it is the directional element that grabs attention. There is something so miraculous about a reference to gravity telling followers how they should in turn use, for example, the position of the sun, that we tend to overlook the more everyday idea of how far to go. But the distance element in the dance is just as important, so how do bees measure out the appropriate movements across the comb – how do the dancers know how far they have flown ? Von Frisch began to look at this as soon as he realised that there was a distance component in the dance. He found that their estimates of the distance to a particular source varied according to the conditions of the journey, so that if they were flying against a wind, they would signal a longer distance than if the wind was behind them. He suggested that this meant that they were using the energy they had expended to estimate how far they had gone. If they used more energy in flying against the wind, they converted that to a greater distance represented in the dance. During the 1950s, a number of experiments were carried out to test the idea, including having bees fly uphill to a food source, so that more energy was needed to cross a given distance. With an echo of the ancient idea that bees picked up small stones to steady themselves in the wind, they were also made to fly with weights, again increasing the energy used in flight. The results were rarely clear, but the energy hypothesis stayed in favour until 1996, when Harold Esch and John Burns in the United States came up with an alternative: optic flow.

Think of yourself travelling in a car. Outside the window, the landscape flashes by. Your eye records it as if it were moving while you were sitting still, and this is optic flow. Distance travelled is measured by multiplying your speed by the time you spend moving. If, inside your car, you recorded the time and you could measure the speed of the movie through the window, you could calculate how far you have gone. Bees experience the same thing, except that they are flying and to them it is the ground that seems to move. The optic flow hypothesis is that bees use their movie, and their sense of time, to estimate how far they have gone. von Frisch himself had considered the idea, but thought it likely to be much less important than measuring the energy used.

Esch and various colleagues now repeated some of the experiments of forty years before. They found that distance signals were unaffected by having to fly uphill, and also by making the bees walk for three metres before they were allowed to fly. Walking uses far more energy than flight, and if the energy hypothesis was correct, the bees which had been made to walk should have signalled a longer distance than they did. They then carried out a new experiment. The food source was now attached to a balloon. At first it was kept on the ground, and then lifted to different heights. The bees were now flying further, and using more energy to climb, so they should have signalled a longer journey, but they did not. So Esch and Burns came back to the optic flow idea and saw a way to test it. Picture yourself travelling again, but this time in a plane. There is no cloud, and you can see the ground, thousands of feet below. Though you are travelling far faster than you ever have in a car, the ground at such a distance hardly seems to move. Optic flow changes with altitude, and the same will be true for bees. So Esch and Burns placed their hive on the roof of the university library, and the food source on the top of another building. They predicted that the bees, which were now higher up and so watching a slower movie of their journey unfold along the ground, would signal a shorter distance than they had actually travelled, and this they did.

Later experiments disrupted optic flow in other ways. Bees can be trained to fly through a tunnel, for example. When they do so, they have to fly close to the walls. We now have the opposite effect to being in a plane, and the 'ground' seems to go by very fast. Their optic flow movie has been speeded up, making it seem to them that they have travelled faster and therefore gone further than they actually have. They signal that the food source is further away than it really is.

If anything, these results should make us marvel even more at how bees communicate with each other. Scientists using dances to see where bees are going have known for a long time that they have to calibrate their bees. And this is because the distance component of the dance varies from place to place – and even within the same hive, for bees travelling north–west rather than south-east, for example. That is, a bee which waggles for 300 milliseconds in a hive in Sheffield is not signalling the same distance as a Brighton bee which waggles for the same time. In fact, it is not signalling a fixed distance at all, but information about a movie. It is saying to its audience: if you fly in the direction that I am showing you, then the length of my dance tells you how much optic flow you need to register in order to go where I have been.

No great distance is needed for the other new pollen type in the last table. The grains are large, about 90 micrometres across. In colour, the pellets are sometimes like the beige of heather, or the amber of maize, but the structure is different from either. They have five, or sometimes four furrow-and-pore combinations, and come from the autumn bedding types of *Viola* that start to appear in gardens now.

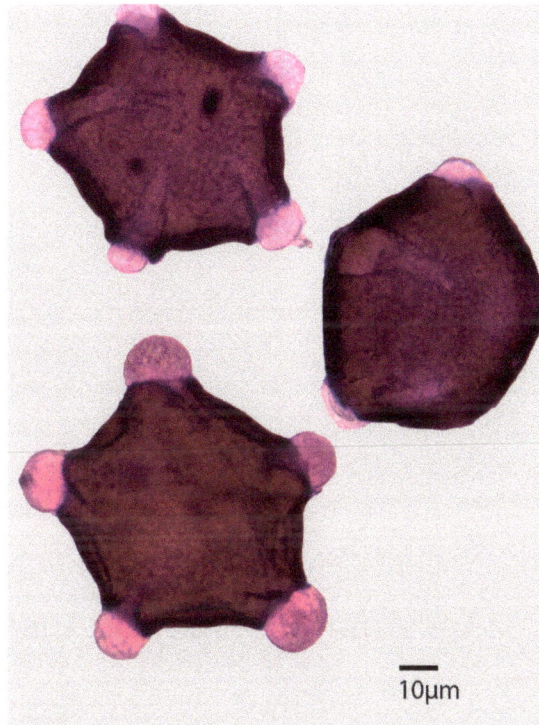

Figure 80: *Viola.* Family: Violaceae (violets and pansies)

Knowing how far bees fly has a number of practical uses. As Huber suggested, it can affect how many hives an area can support. It can also tell us how far we might need to keep bees from something we would rather they didn't eat. Sometimes, for example, honey is marketed as 'organic'. This can only be true if a large enough area is farmed in that way, so that the whole foraging range is within it and the bees have no access to non-organic crops. Or we might be concerned with keeping bees away from plants treated with certain pesticides, or from genetically modified crops. Analysis of bee dances in Germany, in a study which looked particularly at foraging on maize, suggested that maize pollen collection could only be avoided if hives were at least 1500m from the nearest field. The concern with maize was that it would often be treated with neonicotinoid pesticide, and that it flowered at a time of year when the pollen would be being used to raise a particularly important generation of bees: the long-lived workers that take the colony through winter. Two hundred years on from discovering that pollen is used to raise young bees, we are beginning to consider the nature of the diet our landscapes give them.

9. A diet of pollen. August/September 2019.

All other things being equal, we expect bees to gather pollen as close to the hive as they can, and fly only as far as they have to to get as much pollen as they need. But what if other things are not equal: if bees don't always visit the nearest source, but can be tempted further to get something they particularly want ? Eckert, remember, in the 1920s, was surprised to see bees travelling further than seemed necessary, along what he described as a preferred flight lane. As far as pollen is concerned, two possibilities have been put forward to explain why this might happen. There might be differences in the food quality of different pollen types, so that bees sometimes ignore low-quality pollen close to home. And there could be some benefit in variety for its own sake, so that they might have to fly beyond surrounding crops, for example, not because these don't offer enough food, but in order to add different kinds of pollen to their diet.

Although honey bees feed on a great variety of plants, and are quick to adapt to new sources of food, there are plenty that they don't like. 'It is not the farina of every plant that the bee collects,' wrote John Hunter in 1792, 'at least they are found gathering it from some with great industry, while we never find them on others'. The buttercups that were common, but ignored, when the bees were collecting pollen here in the spring are one example, and in this case bees may be avoiding pollen that would do them harm. *Bee World* of 1942 reported that bees in Switzerland would avoid a local buttercup species when there were alternatives, but that bad spring weather sometimes left them with no choice, and then so many could be poisoned that whole colonies would die out. Outbreaks of this 'Bettlach May-sickness' (named after the district in Switzerland) occurred that spring, and in the spring of the year before. Experiments later in 1942 showed that a number of different types of buttercup could produce the symptoms, including the creeping buttercup, *Ranunculus repens,* that is so common in the UK. Honey bees in Britain may be able to ignore it and avoid poisoning because there are many alternative pollen sources in flower at that time of year. Looked at from the plant's point of view, this may be another way in which pollen expenses are kept down. Some species of solitary bees, *Osmia,* use buttercup pollen as one of the main food sources for their larvae, and thrive on it. But other types are poisoned as quickly as the honey bee, and so the buttercup will tend to be visited by only one or two specialists which are able to use it without harm. Since these insects have buttercups largely to themselves, they have no need to visit any other plant, and the buttercup has its own dedicated carrier and less pollen wasted. There is a risk, of course: there are fewer of these insects than honey bees, and they may disappear altogether, but this seems to be the evolutionary gamble that some kinds of plant have taken.

Then there are the other ways in which plants reserve their pollen, which can force honey bees to find something else. For a few weeks now, they have been eagerly visiting the flowers of mallows, whether one of the garden varieties of *Lavatera* or the native Common Mallow that springs up on any patch of waste ground. They come out dusted with white pollen, but they don't pack it into their baskets or bring it back to the hive. In fact, these dusted bees are only looking for nectar, and all they do with the pollen is carry it to the next plant they visit – exactly as the plant would want. Meanwhile, bees which are actually looking for pollen keep away. Experiments with bumblebees have shown that pollen foragers quickly give up with mallow, and so the plant avoids having to give up its pollen for bee food. The reason seems to be the structure of the grains, which are large, and have a surface studded with long spines.

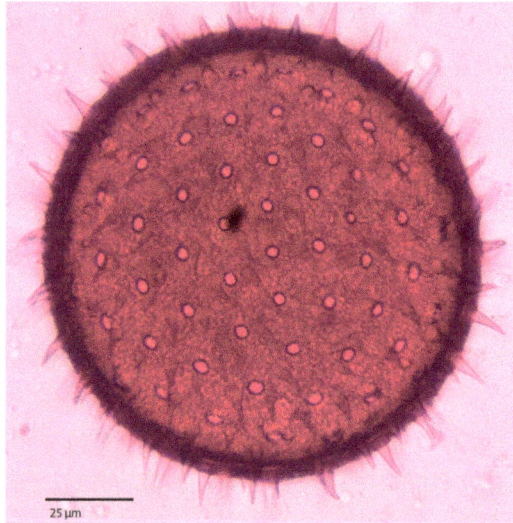

Figure 81: *Malva sylvestris.* Family Malvaceae (mallows)

It may be that the size and spines alone are enough to prevent bees from packing mallow grains into their pollen baskets, although it has been argued recently that other large, spiny pollen grains are gathered, and that there may also be something else, perhaps in the nature of the pollenkitt, that makes this kind unusually awkward to collect. As with buttercups, there are other types of bee, collecting pollen in different ways, that may be more successful, but as far as honey bees are concerned, *Malva* has removed itself from the menu.

Sometimes, then, the pollen diet is controlled by plants. But what about when the bees can choose ? On the one hand, the most important pollen sources in any area will normally turn out to be the most abundant plants, suggesting that honey bees are not fussy feeders, and simply need to collect as much as they can. Their ability to exploit new crops like oilseed rape and maize as these become more common is proof of that, as is the way they have adjusted to the vegetation in any new part of the world to which we have taken them. But then there is some evidence too that the nutritional needs of the colony are seasonal – that the bees need different things at different times of the year – so that foraging may be influenced by this rather than simply reflecting what is most common around the hive. They may then suffer if their choices are restricted by what we plant: experiments carried out around Lancaster found that the protein content of pollen stored in bee bread decreased when there was more arable and horticultural farmland around the hives and increased in landscapes which had more natural grasslands and broadleaf woods. Perhaps, then, not all pollen is the same, and bees are fussier than they seem.

The first part is certainly true: the protein content of pollen can be anything from 2% to 60% of the dry weight of each grain. Different kinds of pollen also vary in the range of protein building blocks - amino acids - that each type can provide. Sunflowers, for example, produce plenty of pollen, but the protein content is relatively low. It also lacks some amino acids of particular importance. Like us, bees are unable to synthesise some of the amino acids that they need, and have to get them directly from their food. They are called the essential amino acids, and for honey bees, it was found in the 1950s that there are ten of them. We can get an idea of their significance from sunflowers. An analysis of different varieties concluded that at least two amino acids were present at such low levels that bees would be unable to meet their needs if they ate sunflower pollen alone. This can be important where sunflowers are the main crop over a wide area, and there is little else that bees can eat. Almonds too, which are often the only food source for the bees moved across the United States to pollinate them in spring, make pollen lacking in several of the essential amino acids. Nor is it just crop plants: dandelions can be a vital source of nectar in early spring, and we have seen the pollen brought back to the hive in March and April this year, but dandelion pollen is deficient in four of the amino acids that honey bees must find in their diet. In experiments where bees were given nothing but pollen from dandelions, these nutritional gaps meant that they were unable to rear brood at all.

This suggests that there are obvious benefits in being able to assess pollen quality before collecting it, but are bees able to do that ? They do seem to be able to assess nectar, and show preference for samples with a higher sugar level. Experiments in the 1950s also suggested that they could assess the balance of sugar types, and preferred solutions with an equal level of glucose, fructose and sucrose. There is some discrimination between pollen types too, at least under experimental conditions. If they are given a choice of pollen samples in the hive, rather than allowing them to forage for themselves, they will often prefer one type to another. Synge, for example, reported a preference for white clover over red, but there is no evidence that this choice is simply based on the protein content of the pollen. Honey bees may be unusual in this: bumblebees seem to be better at choosing pollen with higher protein content and a better range of essential amino acids. Comparison of red mason bees and honey bees foraging on strawberry plants also suggested that the mason bees were more responsive to pollen quality, while honey bees were driven more by the number of flowers open. Perhaps the number of mouths they have to feed make honey bees the least discerning of consumers, with quantity rather than quality their main concern.

It may be, though, that we are looking in the wrong place if we look at what the foragers do. Foragers taste the nectar that they collect, and are in a position to assess it. Pollen, although it must come into contact with their mouthparts and antennae, isn't eaten by the foraging bees at the flower, and so it is hard to see how something like protein content could be assessed. Perhaps we should look instead at the bees which do eat pollen – the nurse bees which process it to produce larval food. What if they assess it and tell the foragers about what they are collecting ? Against this are results showing that nurse bees also seem to be unable to judge the nutritional value of pollen when given a choice in tests. And yet, an image of the honey bee hoovering up pollen regardless of type still seems incomplete, because colonies send out more pollen foragers not only when investigators keep pollen stores down, but also when the protein content of these pollen stores is low. Responses to nutrition have been found as well in colonies kept in enclosures and fed on restricted diets. After a week on a diet lacking an essential amino acid, then provided with a choice, the foragers collected more of the food that offered the amino acid they had missed the week before. These kinds of results would seem to imply that the nurse bees are somehow aware of the quality as well as the quantity of their pollen, and in turn have an influence on what the foragers do.

In addition, there may be components of pollen other than protein which can be assessed, and this time by the foraging bees. As well as protein, pollen also provides the colony with its fatty acids. For bees, they are like protein in two important ways: some can only be obtained from food rather than synthesised in the body, and different types of pollen vary in how much they provide. When bees were fed on pollen deficient in one of the essential fatty acids, then offered a choice between this deficient pollen and one rich in the fatty acid that they now lacked, foragers danced more vigorously for the pollen that would balance their diet. The result of this would be that more followers were recruited to forage at what had become a more valuable source. Unlike protein, it may be relatively easy for foraging bees to assess the fatty acids on offer. Some kinds on the surface of the pollen grains, for example, contribute to the smell of the pollen, and von Frisch showed in the 1920s that honey bees are able to distinguish pollen types by scent.

We can see that the types of pollen that bees collect may then be the result of two factors: what is available in the landscape within range of the hive, and what the bees themselves may choose. Every so often, with a new crop or a new garden plant, we offer them something new, and there is a particularly important example in the last August sample.

August 2019, Week 4

Predominant	*Trifolium*	
Secondary		
Important Minor		
Minor	*Chamaenerion angustifolium*	Rosebay willowherb
	Leucanthemum vulgare	Ox-eye daisy
	Centaurea nigra	Black knapweed
	Rubus	
	Filipendula ulmaria	Meadowsweet
	Fuchsia	
	Melilotus	Melilot
	Taraxacum type	
	Calluna vulgaris	Heather
	Viola	
	Impatiens glandulifera	Himalayan balsam

In amongst the white and beige loads collected from knapweed and heather are pollen loads from *Impatiens glandulifera*. Himalayan balsam is the usual common name, but the fact that it has several, despite being here for less than 200 years, shows what a striking plant it is. With its rapid growth - it is an annual, but can grow two metres or more in its single season - and explosive seed dispersal, it can quickly become the dominant plant in large areas of the wetland habitats it prefers. Because it then replaces native species, it is often vigorously persecuted by conservationists, with work parties organised to pull up the plants before they can set seed. It does have something to offer, though, beyond the exotic appearance that made it a fashionable garden plant for the Victorians: it provides plenty of nectar and pollen, and at a time of year when there are fewer kinds of flowers for insects to find. Like oilseed rape, it is a plant which has quickly become an important source of food. Neither Synge nor Percival recorded *Impatiens* amongst their pollen samples in the 1940s, and Howes described it only as something which 'appears to be a good plant'. He noted casually that it seemed to have become naturalised in a few places, whereas today it grows along virtually every river in the UK. In some areas it is even common enough, like heather, to provide a honey crop of its own. Honey bees climb into the flower to find nectar in the spur at the back, and rub against the stamens above them, so that they often emerge dusted with pollen and come back to the hive a floury white. It looks more difficult for the bee to gather pollen from overhead stamens like this than from something like a poppy, where it can collect the pollen from around its feet, and balsam is not particularly common around my hives anyway, but a few bees manage to work the pollen into pellets and bring it back.

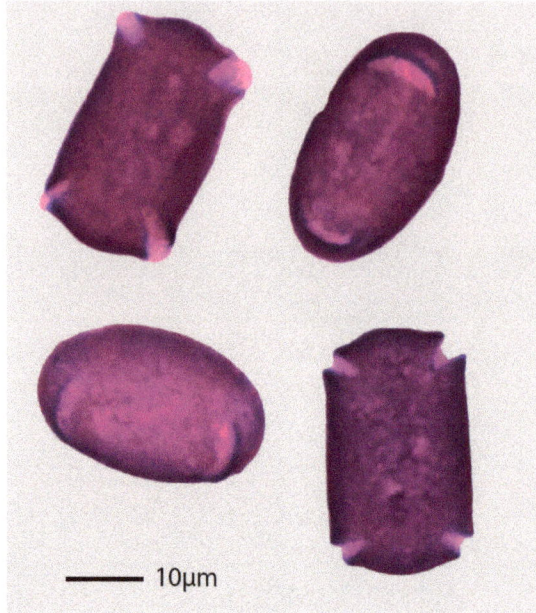

Figure 82: *Impatiens glandulifera.* Family: Balsaminaceae

Fittingly, the exotic invader has unusual grains, with a generally rectangular shape and four apertures instead of the usual three. These are furrows, without pores, and they are quite short, so that grains often look as though they have four nicks cut into the outer coat, one towards each corner.

The 'balsam-bashing' work parties may have had little effect on the spread of *Impatiens*, but methods of biological control are being investigated too, and the tide may yet turn against the invader. No one could really be sorry to see more diverse riverside vegetation return, but beekeepers at least might lose balsam with a few pangs of regret.

Introduced species like balsam can provide a new source of pollen, but equally may replace other sources that were there already. An arable landscape can also provide new crops, but crop fields contain far fewer types of plants than natural vegetation would. Swiss research in an arable area recorded that more than half the honey bee pollen diet came from only five plants, and bees in most parts of the world forage in a landscape where plant diversity is very different from what it would be without human influence. Does this have any effect on honey bees ?

We have already seen that they are capable of using many different types of pollen. Exactly how many is a little difficult to say. All studies which identify the pollen microscopically lump some types together, into a genus like *Prunus* or a whole plant family like the Rosaceae, so the bees may be collecting from more species than the pollen count suggests. If we look at some early results, Synge followed two colonies in 1946, when 37 pollen types were brought to one and 45 to the other. She noted 30 more kinds which were unidentified. Mary Percival also found considerable variety in a single hive, with 86 types collected just between April and August of 1945, 66 of which could be named. The new DNA identification methods may in time improve our ability to separate pollen types: a 2017 study from the University of Würzburg in Germany used them and recorded 149 different types, but this was the total gathered across 16 different hives. It may be that the pollen types collected by each hive were not that different from the results of Percival and Synge, because this was what was found in another study involving multiple hives, published in 2015. This set out 50 hives each year in an arable region of France. The hives were placed in groups of five, and the work was repeated for five seasons. This time, the total number of pollen types was 228 (these were identified under the microscope rather than by DNA) but the average number of pollen types per hive was just over 60. In the hive we are following here through 2019, 61 kinds were found in the trap:

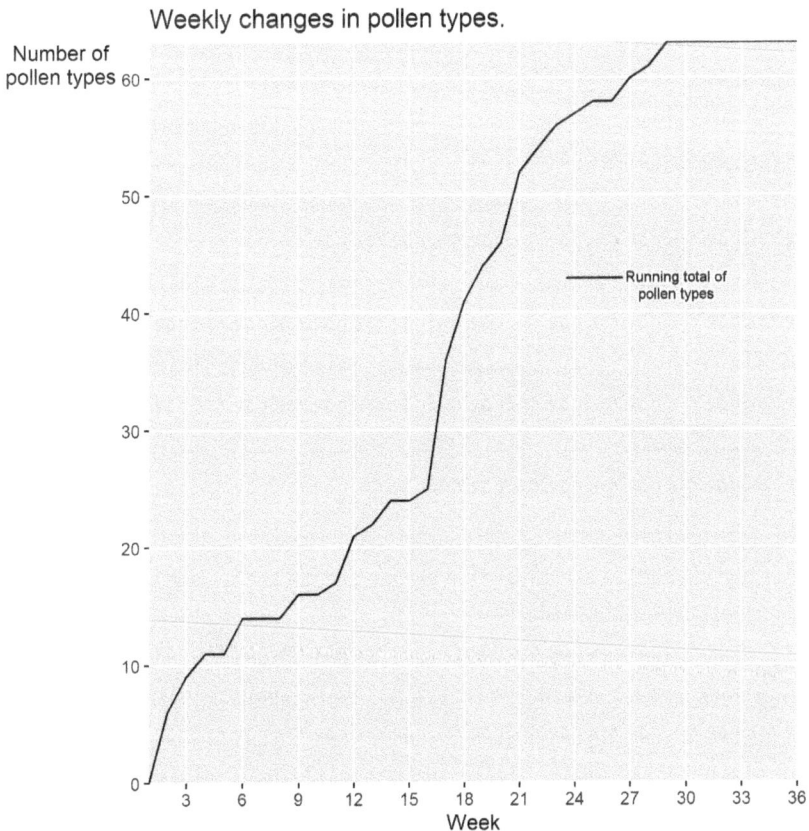

Figure 83: The number of different types of pollen collected in 2019.

The French researchers also looked at to what extent different hives were collecting different types of pollen. As you might expect, the colonies within each group of five tended to collect the same things, having the same plants to choose from. The further apart the hives were, the less their diets resembled each other, which confirms that bees are at least to some extent restricted by what happens to grow around their home. Finally, the general pattern was rather like what we have seen so far in our year of samples: a sequence of predominant types which make up the basic diet, and lots of other kinds gathered in smaller amounts. This seems to be the normal result for honey bees, even in regions which are mostly urban or suburban and may pack an unusually high variety of plants into a small area. Floral diversity reaches its peak in a botanic garden, so the bees studied in the National Botanic Garden of Wales in the spring of 2016 had the widest choice of all. They ignored most of them, and hedgerows and woodland plants were still the main elements in their diet.

Honey bees may not collect as many different types of pollen as they can, but they do seem to gather more than they have to. The French results showed that even when the oilseed rape crop was in full flower, bees were still bringing back pollen from weeds and from patches of woodland on the farms. The Würzburg research was also carried out in a region with lots of oilseed rape, so these bees too were collecting a range of pollen types even in arable areas dominated by an attractive crop. And this was not just when the crop wasn't available. Even when the crop flowered, dancers seemed preferentially to direct foragers to patches of uncultivated land. The result was that hives in areas with less diverse vegetation managed to collect as many types of pollen as hives surrounded by more kinds of plants. The Würzburg researchers concluded that bees in the less diverse areas were flying further in order to find different kinds of pollen. We might conclude from these examples that agricultural land is bad for bees, but this isn't always the case. Researchers in Pennsylvania interested in what made colonies most likely to survive the winter concluded that size had the greatest effect: bigger colonies tended to do best. These were found in areas where more crops were grown and there was less forest, and overall these were the ones with a greater range of pollen and nectar sources.

If this wider range offers an advantage, and if the Würzburg bees are making the effort to fly further for a varied pollen diet, then where exactly is the benefit ? As we have seen, researchers have tended to look for it in the amino acid content. In Poland, though, Michal Filipiak and his colleagues took a different approach, and asked whether we should focus instead on the individual elements from which pollen, and bees, are made. Like all living creatures, bees are built from a wide range of elements: things like the nitrogen and sulphur needed to make proteins, or metals like copper and zinc. Other than carbon, oxygen and hydrogen, a bee gets almost all of these from pollen. The problem, Filipiak suggested, is that the proportions of these elements is different in each type of pollen, and different from the proportions needed by bees. Sunflower pollen, for example, is notably low in phosphorus, so bees might need to seek out another source to have enough of this to build more bees. The explanation for a varied pollen diet might be that this is how bees assemble enough of each of the elements that they need.

If we wander from pollen for a moment to look at another aspect of their diet, looking at the level of the element may also explain something that has long been known about honey bees: their preference for 'dirty' water. Knowing that bees need plenty of water, beekeepers have often provided a fresh source near the hive, and then been disconcerted to see their bees ignoring it in favour of gutters choked with decaying plants, puddles on top of cow pats, or even effluent from sewers. In 1932, Mrs M M Hooper of Cardiff reported in *Bee World* that her bees consistently preferred to drink from a weak solution of salt rather than from pure water, and that they 'are conspicuous for their robust health.' At the same time, C Harrison noted that he had once seen for himself an 'ancient custom' amongst Cornish beekeepers of hanging out a salt pilchard as winter food for bees. These observations, the Polish group proposed, might be explained by noting that pollen in general is lacking in sodium. It may be this, in particular, that the bees are after when they reject the water that looks so much more wholesome to us.

Figure 84: honey bee collecting water

Coming back to pollen, a second advantage to a diverse diet might lie in immunity. In humans, deficiencies in diet can affect the immune system and make us more vulnerable to disease. Insects also have an immune system to respond to infection, and it was reported in 2010 that their ability to do this is reduced when they are fed a less varied protein diet. Because reduced immunity may be one of the factors involved in the higher levels of colony losses that have been recorded over the last decade in some parts of the world, particularly the USA, there has been considerable interest in these results. A further study three years later focused on the nurse bees which consume most of the pollen. When they were healthy, a varied diet had no measurable effect, but when faced with infection by the common gut parasite, *Nosema,* eating more types of pollen meant they lived longer: another hint of a connection between pollen diet and health. Although foraging bees weren't investigated in this study, we know that they are less effective, and less likely to return to the hive, when infected with *Nosema,* so diet might affect their performance too. There is almost certainly no single cause behind increased colony losses: the effect of pesticides and the diseases spread by *Varroa* are other examples in a range of factors that are responsible. But it does seem likely that when we modify the landscape in ways that result in bees having fewer types of pollen to eat, we are adding one more stress to colony life.

September 2019, Week 1

Predominant	*Trifolium pratense*	Red clover
Secondary		
Important Minor		
Minor	Chamaenerion angustifolium	Rosebay willowherb
	Leucanthemum vulgare	Ox-eye daisy
	Centaurea nigra	Black knapweed
	Fuchsia	
	Melilotus	Melilot
	Calluna vulgaris	Heather
	Viola	
	Impatiens glandulifera	Himalayan balsam
	Zea mays	Maize
	Hedera helix	Ivy
	Origanum vulgare	Origanum
	Cirsium	Thistle

As we move through September, fewer plants are in flower, and the bees will have no choice other than to reduce the variety of their diet. In this first week they are still managing to find quite a few types of plant, but only clover in any significant amount. Red clover flowers on later into autumn than white, and was the only clover gathered this week.

Synge, in fact, recorded red clover rather than white as the main type collected throughout August, and the end of her midsummer stage was marked by the end of this clover crop. As a nectar plant, red clover is certainly the less valuable type. The flower heads are larger, and the flower tubes longer, which means that the nectar is often inaccessible to the short tongue of the honey bee. Some of the bumblebees that have longer tongues can use it, but honey bees have to wait until a heavy dew increases the amount of liquid in the nectar tube and brings it within their reach. Then, through capillary action, they can empty the whole tube. Otherwise the plant is useful only for its pollen. The grains have a clear family resemblance to those of white clover, but they are much larger and have a more obvious net pattern on the surface. The larger size also makes it easier to see the granules scattered sparsely along each furrow.

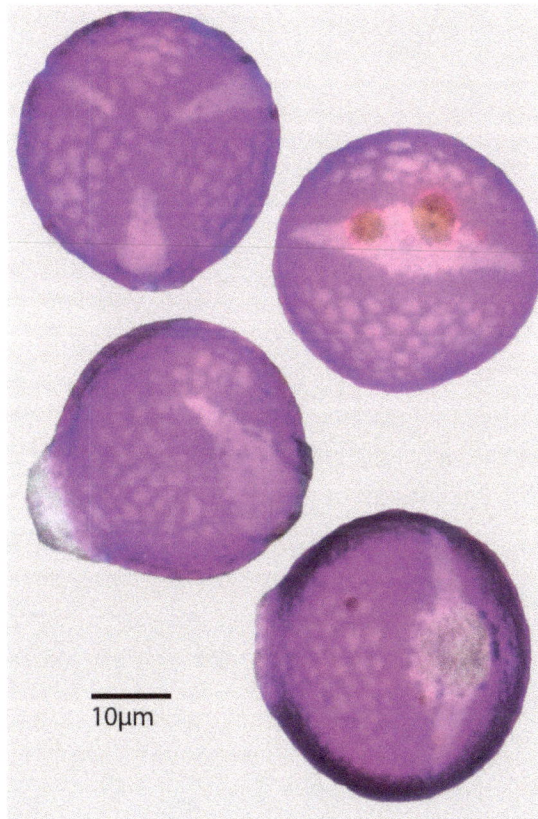

Figure 85: *Trifolium pratense.* Family: Fabaceae (peas and beans)

Amongst the minor types in the last table, there were two new kinds even this late in the year: ivy, which will become the staple of late autumn, and a pollen characteristic of a family we haven't seen so far, the family of mints and sages, with two alternative names, Lamiaceae or Labiatae.

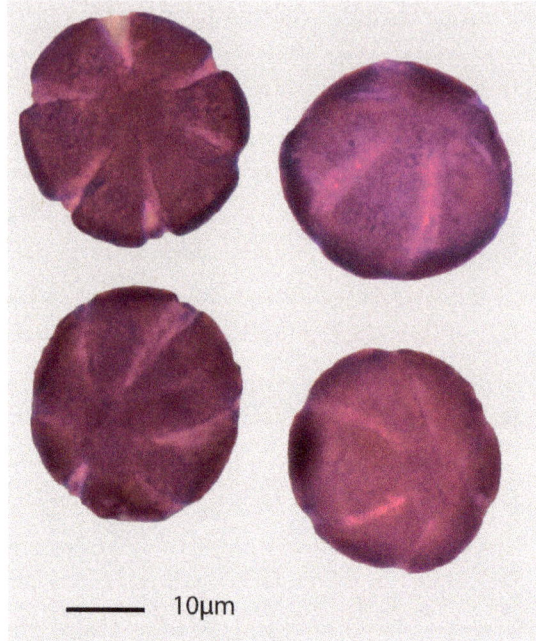

Figure 86: *Origanum vulgare.* Family: Lamiaceae (mints and sages)

Pollen in this family usually has this arrangment of six furrows distributed around the equator, and sometimes a matching hexagonal shape. They are difficult to tell apart, but wild marjoram, *Origanum vulgare,* seeds itself all over the garden and is in flower now, with bees visiting for nectar and collecting pollen too. It seems likely that this is the type in the sample. Like many other plants in its family, marjoram is very attractive to bees, and the structure of the flower is made for insect visitors, with a lower lip on which they can land before pushing their way in for nectar. They would rather these visitors carry the pollen without eating it, though, and pollen is placed on the insect from above, usually in a difficult-to-reach place on its back. Honey bees visiting these kinds of plants usually find the pollen too awkward to collect, and have to be content with nectar. But now, with few pollen sources about, one or two seem to have adopted the acrobatics needed.

It seems, then, that honey bees seek out a varied pollen diet when they can, but without assessing each pollen type as they find it. This lack of assessment may be why they will sometimes collect harmful pollen like that of buttercups, or of castor beans, widely grown in the tropics for biodiesel. It also means that we have the potential to harm them with the abundant pollen that our crops sometimes provide. Maize, where pollen collection is coming to an end now, illustrates the two main concerns. The first is genetic modification, banned in Europe for crops, but widespread elsewhere in the world. Maize of this kind, first grown commercially in the US in the late 1990s, is often called Bt maize, because it contains genetic material from a soil bacterium, *Bacillus thuringiensis.* These genes allow the maize to produce toxins which kill the larval butterflies and beetles that are significant pests on the crop. Because the toxin is already inside the plant, the grower no longer has to handle insecticides, and they don't have to be sprayed across the crop. Insecticide 'drift' from spraying is often a problem for apiaries near to farmland, so reducing this is helpful, but if honey bees collect the pollen of Bt maize, their larvae will still be exposed to the toxins that the plant's new genes have made. Early reports that the larvae of the Monarch butterfly can be poisoned by modified maize pollen fuelled concern that other non-target insects might be affected, but bees are more distantly related to the pests, and repeated studies over the last twenty years have shown no effects. Even if Bt maize comes to Europe, it seems that, as far as the bacterial toxins are concerned, its pollen will be safe.

There is another way in which crops are protected, though, which may do more harm. Here the seeds are already coated with insecticide, and as the plant grows the insecticide travels through it, protecting each part of the plant but then finding its way also to the nectar and pollen which bees collect. In the late 1990s, a new class of insecticide that could be applied in this way was introduced, and soon became almost universal on, amongst other crops, maize and oilseed rape. These were the neonicotinoids, which act on the nervous system of insects, and they were marketed as having less harmful effects on wildlife than previous generations of insecticide. Soon, though, there were concerns that this might not be true for bees.

There is no doubt that the chemical is present in nectar and pollen: the argument became over whether the amounts present were large enough that they could have any effect. Pests like the flea beetle which attacks oilseed rape would be poisoned by eating the plant itself, but perhaps bees feeding on small amounts of nectar or pollen would come to no harm. Testing this proved controversial. On the one hand, it seems that bees can be fed a dose of insecticide equivalent to what they would ingest what foraging (and there were many arguments about what a realistic dose might be) and not die. On the other, might the insecticide accumulate in the insect over its foraging career, and over time have a different effect ? Or, and this became the key point as far as bees were concerned, might an insecticide which targeted the nervous system affect their ability to learn about their environment, find food and their hive, and communicate with the colony ? In this case, measuring the dose that made the bee drop dead was irrelevant: it was the effect on behaviour that mattered, and this was far more difficult to test. Effects were reported on other bees too: reduced reproduction in bumble bees and solitary bees, although there were also studies which showed no effects.

The potential for harm, opponents of the neonicotinoids argued, was shown by the mass poisoning of honey bees that occurred with maize crops in Germany in 2008. A government investigation concluded that incorrect practice had exposed bees to high levels of a type of neonicotinoid and affected almost 11,000 colonies. Without admitting liability, the manufacturers paid 2m into a fund set up to compensate beekeepers in the region. Direct poisoning like this was unusual, but it helped spur lobbying against the chemicals. Five years later, use of the insecticide in question, along with two other kinds, was suspended in the EU for use on crops attractive to bees. In the same year, beekeepers and environmental groups in the US sued their Environmental Protection Agency for the harm that they claimed had resulted from approving the use of neonicotinoids. In 2018 the EU ban became permanent, and was extended to cover almost all crops grown outdoors. By then, large-scale field studies on maize in Canada, and oilseed rape in sites in Hungary, Germany and the UK, had concluded that neonicotinoids were having a negative impact on the three kinds of bee that they looked at: honey bees, a species of bumblebee and a type of solitary bee. One of the main sources of the chemicals was pollen. The bees were not being directly poisoned, but receiving enough of the insecticide, perhaps in combination with other environmental factors, to reduce survival from one season to the next. The UK, which had opposed the earlier suspension, changed its position and supported the ban.

Other types of neonicotinoids can still be used, and the manufacturers and some farmers' groups continue to campaign to have the ban repealed. Some argue that growers will have to go back to older insecticides, with worse effects on wildlife, or that it will no longer be economic to grow oilseed rape at all. Other responses to the ban are more hopeful, though, and some farmers in France and England have had good yields from mixing their rape with companion crops like field beans and buckwheat and so diverting the attention of the pests. As well as reducing losses to the flea beetle, this approach might help bees too, by adding to the variety of food plants they can use. As Robert Huish put it, our fields may yet be places 'in which they love to disport'.

10. The end of the year. September/October, 2019.

Looking at her results from 1945 and 1946, Synge defined the end of the pollen-gathering summer as the point when her bees stopped bringing red clover pollen to the hives. After that, for a month or so from the first half of September, was the autumn stage, the last of the year.

September 2019, Week 2

Predominant	*Trifolium* mix	Clover
Secondary		
Important Minor	*Hedera helix*	Ivy
	Calystegia sepum	Hedge bindweed
Minor		

Synge's bees collected a little autumn pollen from some garden sources: dahlias, nasturtiums and Japanese anemones among them; more from the white mustard grown as a green manure crop around her study site, but ivy was the main source for the final month. Ivy was also the last pollen source of the year in the Irish study of Coffey and Breen in the 1990s. The pattern here is the same, with a few pellets of Fuchsia pollen still being collected, late Brassica flowers offering the last field crop source, but ivy gradually becoming the dominant type.

September 2019, Week 3

Predominant		
Secondary	*Trifolium* (most repens)	Ivy
	Hedera helix	
	Brassica	
Important Minor		
Minor	*Fuchsia*	

There was rain all the next week in 2019, and no pollen collected. But there are usually fine days in autumn too, and a clump of flowering ivy in the sun bustles with insects: butterflies, hoverflies, the last wasps, and honey bees, many of them collecting its orange-yellow pollen. There is even a solitary bee which emerges late in the year and uses ivy pollen as its main source of food. This ivy bee, *Colletes hederae*, has only been recorded in Britain since 2001 and seems gradually to be spreading north. The pollen which it collects, mixed with nectar, is stored in burrows, and next year's larvae will feed upon it when they emerge. Some of the grains collected by honey bees now will also be stored in the hive to feed the larvae next spring. Under the microscope, this last pollen type could have been designed to form the most typical of grains, showing all the features that have become familiar over the year. They are round, or three-cornered, depending on how they lie. They are of average size, somewhere between 25 and 30 micrometres across. They have three apertures, each a furrow with a central pore. The surface is patterned with a net. The grains are larger than those of willow, but they are similar otherwise in form, and in the orange, green and yellow shades of the pollen loads. As the pollen year comes to an end, the last main crop recalls the first.

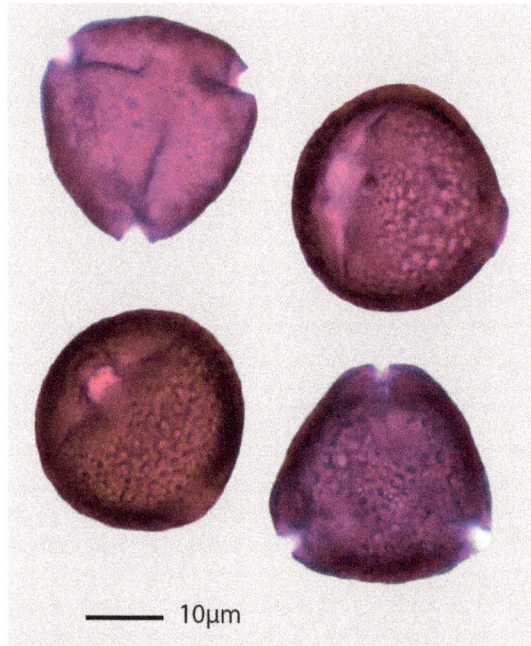

Figure 87: *Hedera helix.* Family: Araliaceae.

October 2019, Week 1 (two days)

Predominant	*Hedera helix*	Ivy

October 2019, Week 2 (one sample day)

Predominant	*Hedera helix*	Ivy

October 2019, Week 3 (one sample day)

Predominant	*Hedera helix*	Ivy

October 2019, Week 4 (two days)

Predominant	*Hedera helix*	Ivy

There were lots of days in October when rain or cold kept the bees inside, but enough fine weather for at least one pollen sample each week. Ivy was the only kind trapped now until the pollen season ended at last in the first week of November. For at least a month at the end of the season, honey bees forage almost exclusively on ivy, and its importance as the colony prepares for winter has been known for a long time. Howes wrote that it was sometimes possible even to take a honey crop from it, and that the honey had a good flavour: this in spite of the unpleasant scent of the flowers which, he thought, might account for the carrion flies usually found amongst the butterflies and bees. A recent report from the Social Insects laboratory in Brighton confirmed its significance. Its pollen accounted for an average of 89% of the autumn collection, and it was even more important for nectar, with 80% of its honey bee visitors only collecting that. What makes ivy particularly useful is that it is such a widespread plant: the Brighton study looked at rural and urban habitats and found it to be common in both. It would be unusual for any colony not to have patches of ivy flowering nearby.

Each flower provides pollen first, with bright yellow anthers to help attract bees, and then the anthers are lost and nectar is produced. The nectar is unusual. Hermann Müller noted in 1878 that so much was made that a white sugary crust would form around the nectary if the nectar wasn't gathered by insects. This suggests a high concentration of sugar, and ivy nectar is valuable partly because of this. It can be as much as 49% by weight according to the Brighton work, which they compared with other important nectar plants like brambles (15-39%) and white clover (23 – 34%). This means that there is less water to be evaporated before the nectar can be stored, which is an advantage in the cooler autumn weather, when evaporation will be slower. But the rapid crystallisation that Müller reported is also caused by the unusual balance of sugars in ivy nectar, and this can create problems. The sugars in nectar are mainly a mixture of fructose, glucose and some sucrose. The balance between them varies with different species: one of the reasons why clover and melilot are so attractive to bees seems to be that their nectar contains almost equal levels of the three sugar types. Of the 889 kinds of pollen examined by Mary Percival in 1961, only ivy had a considerable excess of glucose: confirmed by other studies as forming 75 – 80% of the nectar sugar content in this plant. Glucose forms crystals more readily than fructose, and honey made from a glucose-rich nectar will granulate much more quickly. This is why beekeepers have learned to extract oilseed rape honey quickly, before it crystallises in the hive: brassica nectar also has a higher than usual ratio of glucose to fructose. Ivy honey, where the ratio is higher still, will do the same, and a commercial beekeeper in Ireland reported in 1974 that he had lost several colonies over the previous winter, following autumn feeding on ivy and granulation of the honey stores. Some of the combs were examined at Rothamsted, and their conclusion was that a shortage of water may have been the problem. In its liquid state, honey is water saturated with sugars. When solid crystals form, the water left behind now forms a more dilute sugar solution. Under some circumstances, that can then ferment, but in this case, in the warm conditions of the hive, so much of the water may have evaporated that the bees were no longer able to feed on the sugar crystals, and starved. This was an unusual event, though, and normally ivy flowers are an important resource for bees and many other insects too.

They certainly form the last significant food source, of either nectar or pollen. In my garden, bees can still emerge on any sunny winter day. Some visit the flowers of Fuchsia 'Lottie Hobby' that will persist into January if there is no heavy frost, but they are only there for nectar. The pollen year is over until the spring.

At this point we can look back and see the pattern of the year. With colours to represent the categories, from predominant to minor, the whole year looks like this:

A summary of pollen collection in 2019

	February				March				April				May				June				July				August				September				October			
	1	2	3	4	1	2	3	4	1	2	3	4	1	2	3	4	1	2	3	4	1	2	3	4	1	2	3	4	1	2	3	4	1	2	3	4
Corylus avellana																																				
Galanthus nivalis																																				
Viburnum tinus																																				
Alnus glutinosa																																				
Crocus																																				
helleborus orientalis																																				
Taxus baccata																																				
Ulmus																																				
Tussilago farfara																																				
Salix																																				
Prunus group																																				
Buxus sempervirens																																				
Taraxacum officinale																																				
Skimmia japonica																																				
Ficaria verna																																				
Hyacinthoides non-scripta																																				
Brassica																																				
Acer pseudoplatanus																																				
Quercus																																				
Fagus sylvatica																																				
Aesculus hippocastanum																																				
Crataegus monogyna																																				
Plantago																																				
Clematis montana																																				
Sambucus nigra																																				
Meconopsis cambrica																																				
Chenopodium																																				
Taraxacum type																																				
Dianthus barbatus																																				
Ceanothus																																				
Papaver rhoeas																																				
Trifolium																																				
Rosa																																				
Rhododendron																																				
Heracleum spondylium																																				
Geranium																																				
Polemonium caeruleum																																				
Cornus																																				
Rubus																																				
Rumex																																				
Symphytum																																				
Ligustrum vulgare																																				
Graminae																																				
Hypericum																																				
Castanea sativa																																				

Legend:
- predominant
- secondary
- Important minor
- minor

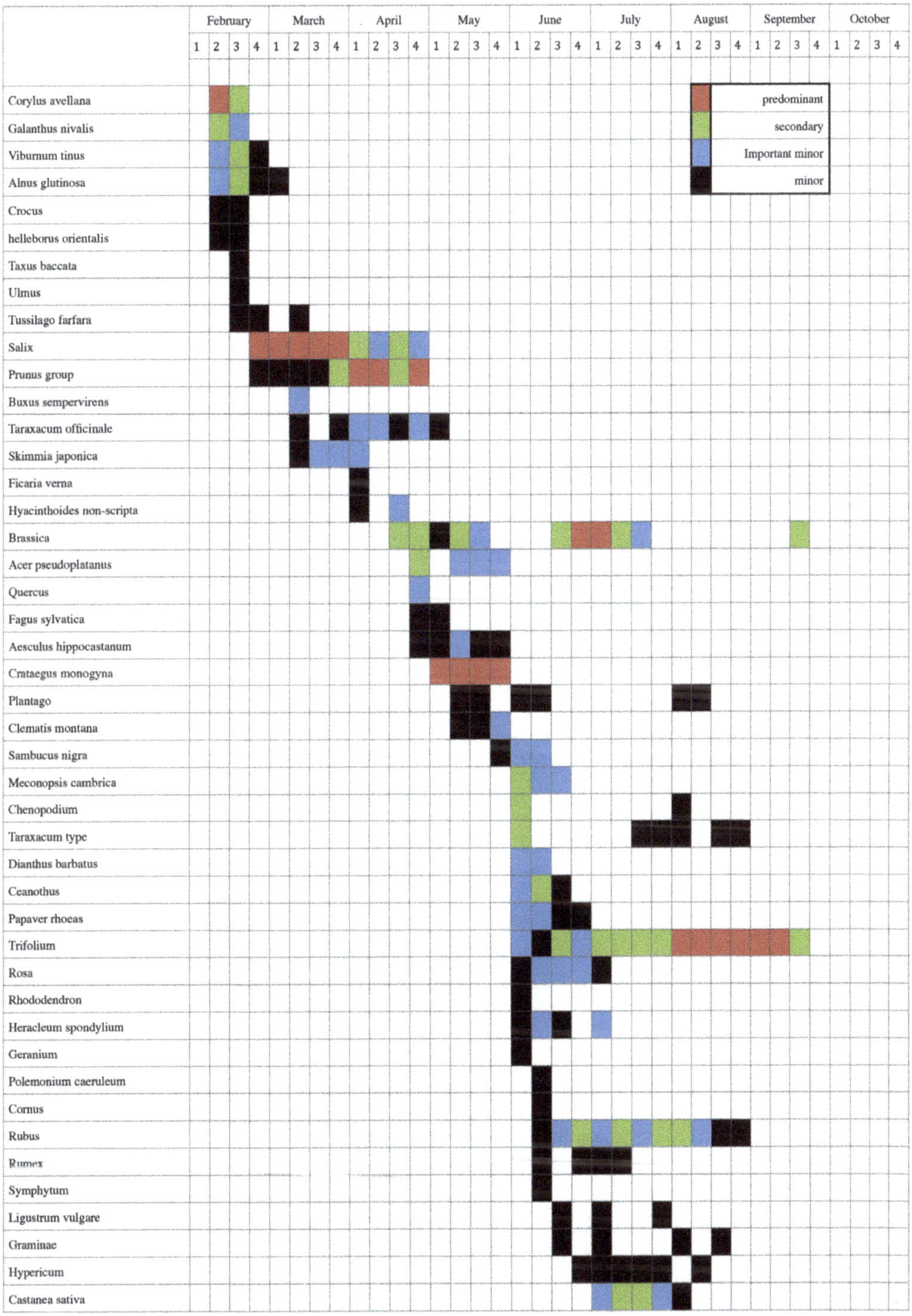

Figure 88: Pollen collection in 2019.

Filipendula ulmaria	
Calystegia sepium	
Cirsium	
Chamaenerion angustifolium	
Melilotus	
Centaurea nigra	
Leucanthemum vulgare	
Fuchsia	
Hydrangea	
Clematis vitalba	
Zea mays	
Calluna vulgaris	
Viola	
Impatiens glandulifera	
Origanum vulgare	
Hedera helix	

And if we then draw out the details, we could look first at the red squares of the predominant types. Here a steady succession of these suggests that in this landscape, the colony is rarely without a good source to form the staple of their pollen diet over the weeks that it flowers. Without a brassica crop, though, probably mostly oilseed rape, there would be a significant gap in June and July. Then a variety of other sources is needed to keep the colony supplied.

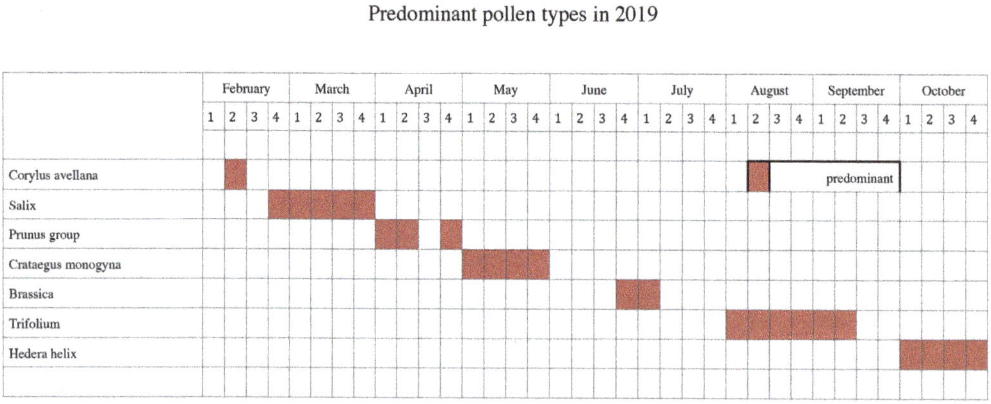

Predominant pollen types in 2019

	February				March				April				May				June				July				August				September				October			
	1	2	3	4	1	2	3	4	1	2	3	4	1	2	3	4	1	2	3	4	1	2	3	4	1	2	3	4	1	2	3	4	1	2	3	4
Corylus avellana																										predominant										
Salix																																				
Prunus group																																				
Crataegus monogyna																																				
Brassica																																				
Trifolium																																				
Hedera helix																																				

Figure 89: Predominant pollen types.

Secondly, we can examine variety by looking at the vertical blocks of colour in figure 85, which show how the number of pollen types collected changes from week to week. In the first week, for example, February 2, there are six pollen types in the column, from *Corylus* to *Helleborus*. A graph of this information over the season is partly just a reflection of how the number of flowering plants changes during the year, but is also a reminder of some of the questions we still have about the pollen diet. The arrows show how the number of types collected sometimes falls when an important plant like willow comes into flower. We can picture the colony concentrating its foragers on this abundant source. But at other times, as with clover, the variety of the diet remains high, as if there were, as we have discussed, some advantage in collecting different types.

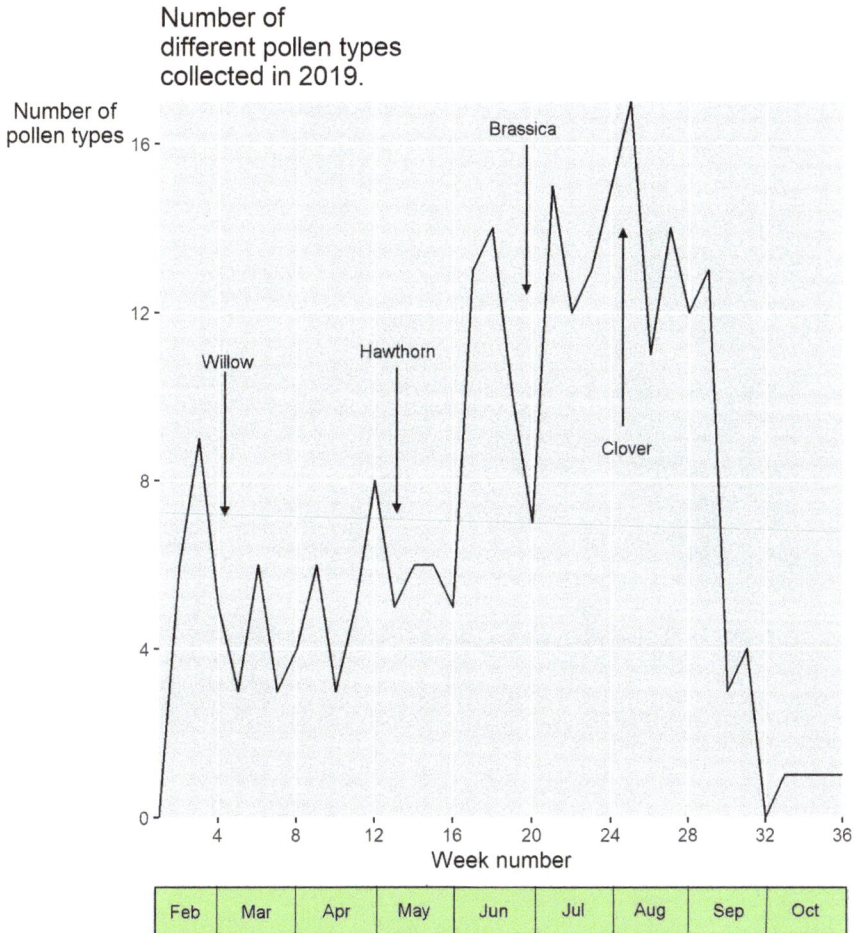

Figure 90: Number of pollen types collected each week.

Some kinds of pollen, like willow or clover, are predominant types over several weeks, and are clearly important components of the pollen economy of the hive. Many other kinds are only minor sources for a short time. We can use these two attributes of abundance and time to score each type of pollen with a simple measure of its importance. So if, for example, clover has a score of 4 for each week it is a predominant source, 3 each time it is secondary, 2 for being an important minor source, and 1 when it is minor, its overall score for the year is 47. And if we do this for all 61 types in 2019, we can see how 30 have a score of less than 5, and 50 a score under 10. The diet is a few staples, along with a rich variety of types collected in small amounts.

Distribution of pollen scores.

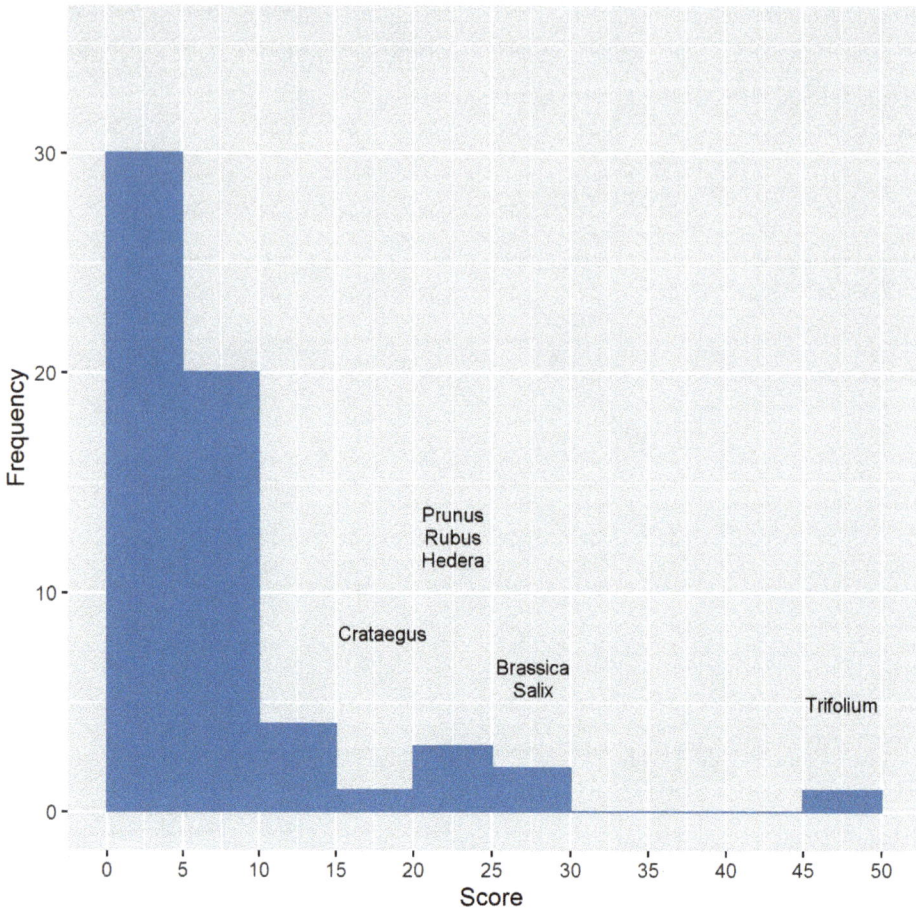

Figure 91: Distribution of pollen scores.

Then if we look at the types with a score of ten or more, we can see that the main elements in the pollen diet are these:

Pollen types with a score of ten or more.

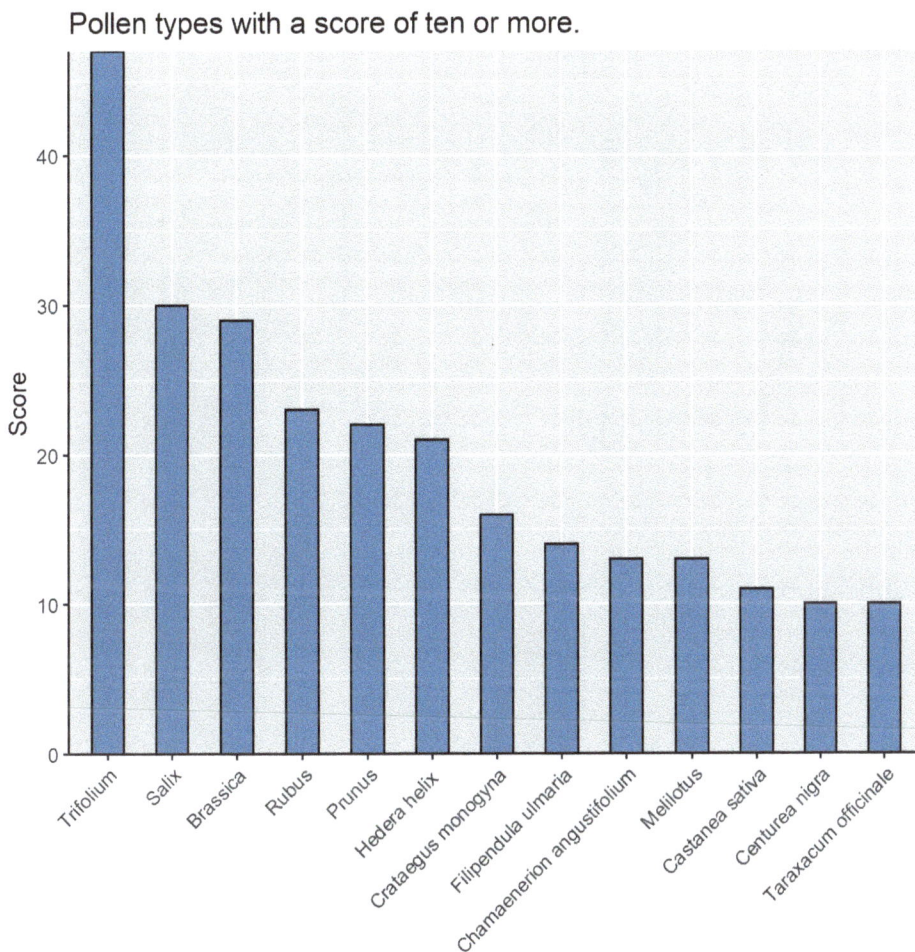

Figure 92: Pollen types with a score of ten or more.

Some of these, like meadowsweet and melilot, never reach the predominant class, but they are collected every year and over several weeks. They contribute almost as much as hawthorn, which is a predominant type this year whenever it is collected (Mary Percival once recorded more than 21,000 hawthorn loads on a single day) but only flowers for a month. Other than hawthorn, though, there is a clear step from the predominant types to everything else, and the list of these main pollen plants is a fairly short one. It is also, in spite of the changes of the last 70 years, a list that Synge and Percival would recognise. Four of the five most common types in Synge's study are the same as they are here, if we exchange oilseed rape for sainfoin as a source of brassica pollen. Percival had charlock *(Sinapis arvensis)* as a brassica, and her season was too short to include willow and fruit trees in spring, and then ivy in autumn, but otherwise her pattern too is much the same.

Mary Percival: Main pollen types in 1945.

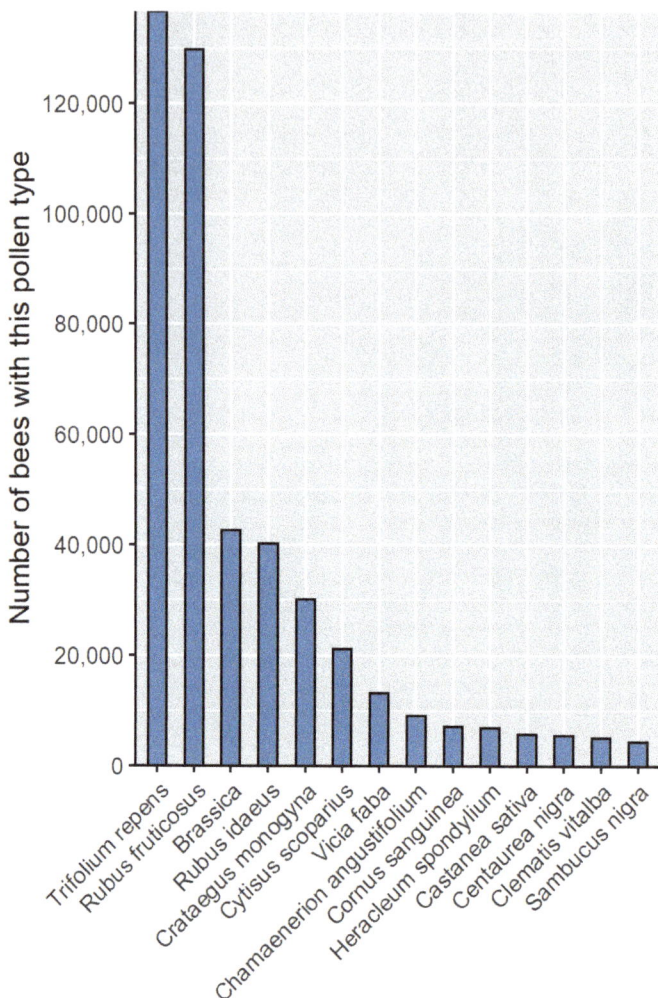

Number of bees with this pollen type

120,000
100,000
80,000
60,000
40,000
20,000
0

Trifolium repens
Rubus fruticosus
Brassica
Rubus idaeus
Crataegus monogyna
Cytisus scoparius
Vicia faba
Chamaenerion angustifolium
Cornus sanguinea
Heracleum spondylium
Castanea sativa
Centaurea nigra
Clematis vitalba
Sambucus nigra

Figure 93: Percival's main pollen types, 1945.

There are parts of Britain where the landscape has changed greatly since these early pollen studies, and the pollen diet of honey bees may be quite different. For my bees, there are more houses and some new roads. Industrial decline has created brownfield sites, often temporary but for a while rich in pollen and nectar plants. But farming within the colony's foraging range has changed less than in many parts of lowland Britain, with only a small area joining the boom in oilseed and maize. Some of the grassland is more intensively managed, some can have hardly changed.

Within this particular landscape, each component has something to contribute. Gardens and road verges contain a good number of the minor sources of pollen, and are also home to many of the fruit trees that are the main element of the diet in late spring, and the ivy which is so important at the end of the year. And if this study agrees with the continental surveys that have shown the significance of woodland and other uncultivated ground, it also emphasises the value of fields in which clover can grow and there is hawthorn in the hedges. The pollen season for honey bees is a long one, and can only be filled by a mixture of habitat types.

Of course, there is no particular reason to shape our landscapes around honey bees. Some even suggest that, like the red squirrel or giant panda, they attract more attention than they should while other creatures go unnoticed. But the bees and their pollen year have two messages which go beyond a single kind of insect. Firstly, if we can create places where even these resilient and adaptable creatures can't thrive, then that should alarm us, meaning that we could be past the stage where the canary chokes, and about to watch the miners fall. And secondly, landscapes where they can thrive are likely to offer something for plenty of less-obvious species too. And even for us. If the pollen year requires a variety of habitats and as long a season of flowers as possible, does that not sound like the kind of landscape in which we would like to live ? To give the last word to disgrunted and disreputable Robert Huish, a place in which the honey bee loves to disport will surely delight our eyes and gladden our hearts too.

Index of pollen images

Notes

Although not intended as a full list of references, these notes may be of use to readers who want to follow up some of the sources used.

Chapter 1.

1. The COLOSS project, of which the pollen work is a part, can be found at https://coloss.org/ and the honey monitoring scheme at https://honey-monitoring.ac.uk/

2. R.O.B. Manley, *Honey Farming*. (Reissued 2012, Northern Bee Books).

Chapter 2.

1. Percival M (1947), Pollen Collection by *Apis mellifera*.
The New Phytologist 46:142-173.

2. Pollen trap effects are discussed by Hoover & Ovinge (2018), Pollen Collection, Honey Production, and Pollination Services: Managing Honey Bees in an Agricultural Setting, *Journal of Economic Entomology* 111(4):1509-1516. Plans for the OAC pollen trap are available online: https://www.ontariobee.com/sites/ontariobee.com/files/document/construction.pdf

Figure 3, of course, shows the trap pulled out from the hive.

Chapter 3.

1. Echlin P(1968). Pollen. *Scientific American* 218 (4) (April 1968).

2. Farrar CL (1937), The Influence of Colony Populations on Honey Production.
Journal of Agricultural Research 54:945-954. Available online at https://naldc.nal.usda.gov/download/IND43969007/PDF

3. Coltsfoot pollination: Wild J & Gottsberger G (2003), Pollination and reproduction of Tussilago farfara (Asteraceae). Botanische Jahrbücher für Systematik, Pflanzengeschichte und Pflanzengeographie 124:273 – 285.

4. Allelopathy in dandelion pollen: Loughnan D et al (2014), *Taraxacum officinale* pollen depresses seed set of montane wildflowers through pollen allelopathy.
Journal of Pollination Ecology 13:146-150.

5. Arthur Dobbs: Grant V (1949), Arthur Dobbs (1750) and the Discovery of the

Pollination of Flowers by Insects, *Bulletin of the Torrey Botanical Club* 76(3):217-219.

6. Synge AD (1947), Pollen Collection by Honeybees, *Journal of Animal Ecology* 16(2):122-138.

7. Coffey M & Breen J (1997), Seasonal variation in pollen and nectar sources of honey bees in Ireland, *Journal of Apicultural Research,* 36:63-76.

8. Wisconsin pollen: Severson, D. W. &Parry, J. E. (1981). A chronology of pollen collection by honeybees. *J. Apic. Res.* 20, 97–103.

Chapter 4.

1. There is a introduction to Huber's *New Observations on the Natural History of Bees* (1792) at http://www.bushfarms.com/huber.htm.

2. Alexander Deans published his *Survey of British Honey Sources* as a Bee Research Association pamphlet in 1957.

Chapter 5.

1. Percival M (1950). Pollen Presentation and Pollen Collection. *New Phytologist* 49:40-63.

2. Casteel's description, *The Behavior of the Honey Bee in Pollen Collection* can be found at http://www.gutenberg.org/ebooks/40802. Sladen (1912), How Pollen is Collected by the Honey-Bee, *Nature* 88, 586–587.

3. Ribbands CR (1949), The Foraging Method of Individual Honey-Bees. *Journal of Animal Ecology* 18: 47-66.

4. R.L.Parker's account of pollen collection (1926) has been digitised by Cornell University at http://reader.library.cornell.edu/docviewer/digital?id=chla7251474_8558 _006#page/10/mode/1up

5. There is an excellent introduction to the bee bread argument at http://scientificbeekeeping.com/tag/dr-anderson/

6. France: Requier et al (2015), Honey bee diet in intensive farmland habitats reveals an unexpectedly high flower richness and a major role of weeds. Ecological Applications 25:881-890.

7. Poppies: McNaughton I & Harper JL (1960), The Comparative Biology of Closely Related Species Living in the Same Area. I. External Breeding-Barriers Between Papaver species. *New Phytologist* 59:15-26.

Chapter 6.

1. von Frisch: https://www.mpg.de/789351/W006_Culture-Society_074-080.pdf.

2. Round dances: Griffin S, Smith M & Seeley T (2012), Do honeybees use the directional information in round dances to find nearby food sources? *Animal Behaviour* 83:1319-1324.

3. Hasenjager M, Hoppitt W & Leadbeater E (2020), Network-based diffusion analysis reveals context-specific dominance of dance communication in foraging honeybees, *Nature Communications* 11:625.

4. An introduction to some of the dance language arguments is Dyer F (2002), The Biology of the Dance Language, *Annual Review of Entomology* 47:917-949.

Chapter 7.

1. Spores: Zabriskie, J.L. (1875) Remarkable forage for bees. Beekeepers Magazine 3: 186–187. Lang WH (1901) Fungus spores as bee-bread. *The Plant World* 4:49-51.

2. Howes FN (1945) *Plants and Beekeeping.* Faber & Faber, London.

3. Visscher PK & Seeley T (1982) Foraging Strategy of Honeybee Colonies in a Temperate Deciduous Forest. *Ecology* 63:1790-1801.

4. Landscape and dance: Nürnberger F, Steffan-Dewenter I & Härtel S (2017) Combined effects of waggle dance communication and landscape heterogeneity on nectar and pollen uptake in honey bee colonies. 10.7717/peerj.3441. Couvillon M, Schürch R & Ratnieks F (2014) Dancing Bees Communicate a Foraging Preference for Rural Lands in High-Level Agri-Environment Schemes. *Current Biology* 24:1212-1215.

5. Worcestershire: Adams-Groom B, Martin P, Sierra-Banon A-L. 2019. Pollen characterization of English honey from Worcestershire, West Midlands (UK). Bee World, 97 (2): 53-56

6. Pollination efficiency: Cruden R & Jensen K (1979) Viscin Threads, Pollination Efficiency and Low Pollen-Ovule Ratios. *American Journal of Botany* 66:875-879. Song Y-P, Huang Z-H & Huang S-Q (2018) Pollen aggregation by viscin threads in Rhododendron varies with pollinator *New Phytologist* 221:1150-1159.